Emulsions and Solubilization

Emulsions and Solubilization

KŌZŌ SHINODA
Department of Applied Chemistry
Yokohama National University
Yokohama, Japan

STIG FRIBERG
Department of Chemistry
University of Missouri
Rolla, Missouri

A WILEY-INTERSCIENCE PUBLICATION
JOHN WILEY & SONS
New York Chichester Brisbane Toronto Singapore

Library of Congress Cataloging in Publication Data:

Shinoda, Kōzō, 1926-
Emulsions and solubilization.

"A Wiley-Interscience publication."
Includes bibliographies and index.
1. Emulsions. 2. Solubilization. 3. Surface active agents. I. Friberg, Stig, 1930- II. Title.

TP156.E6S55 1986 660.2′94514 86-5495
ISBN 0-471-03646-3

Printed in the United States of America

10 9 8 7 6 5 4 3 2 1

Preface

When we started the research on solubilization and emulsions in 1963–1967, we felt the need to study the temperature change of a system containing nonionic surfactants against the fact that with temperature elevation, the solubilization in nonionic surfactant solutions changes in a pronounced manner and that emulsions often change their type from oil/water (O/W) to water/oil (W/O). At that time, surprisingly enough there was practically no information on the temperature effect in nonionic surfactant solutions in authoritative books on emulsion properties. This motivated us to study the properties of emulsions stabilized by a nonionic surfactant of the type polyethylene glycol alkyl(aryl)ethers. From the studies of emulsion, solubilization, and related phenomena as a function of temperature, we have also acquired a fairly complete knowledge of the effect of the hydrophilic chain length of nonionic emulsifiers on various properties, since temperature rise in nonionic surfactant solution corresponds to a shortening of the hydrophilic chain length of surfactant. It is the usual practice for surfactant scientists to select a suitable nonionic surfactant by finding the optimum hydrophilic chain length at a given temperature. With the parallel behavior of temperature change and alteration of polar chain length established, many figures on the temperature effect in this book will be very useful and suggestive.

On the other hand, the temperature effect on ionic surfactant solutions is small and the size of hydrophilic groups is fixed. Hence, the control of the hydrophile–lipophile balance (HLB) of the ionic surfactant is not as directly achieved as with the nonionic surfactants. Several indirect devices are used:

1. Bivalent salts of ionic surfactants soluble in hard water;
2. More lipophobic, oil soluble, cosurfactants;
3. Well-balanced ionic surfactants.

Hence, the HLB, that is, the solution property of the ionic surfactant, may continuously change with the salt concentration or its composition from water soluble to oil soluble. At respective optimum HLBs of surfactant (or mixture) the solubilization is maximum, oil/water interfacial tension is minimum, and emulsions are most stable.

The purpose of this book is to provide information to emulsion and surfactant scientists on how logically and readily solubilization, microemulsions, and emulsion stability or inversion, and so on, are understood as a function of the HLB of surfactant. This book is suitable for use as a textbook. It will help the reader to understand underlying phenomena and enhance his or her potential for solving problems.

Kōzō Shinoda
Stig Friberg

Yokohama, Japan
Rolla, Missouri
June 1986

Acknowledgments

We would like to thank all our teachers, students, collaborators, and friends for intellectual stimulus over the years, our families for their understanding attitude, and Ms. Geritzi Forren for patiently and competently typing our manuscript over and over again.

K.S.
S.F.

Contents

Introduction 1

I.1 Definition of Emulsions 3

I.2 Definition of Solubilization 4

I.3 Definition of Microemulsion 5

I.4 Definition of HLB, HLB number and HLB Temperature (Phase Inversion Temperature = PIT in emulsion) 6

References 7

1 Correlation Between Solution Behavior of Surfactants and Solubilization Microemulsions or Emulsion Types in Surfactant/Water/Oil Systems 11

1.1 The Effect of Temperature and/or the Hydrophilic Chain Length of Emulsifier on the Solution Behavior of Nonionic Surfactant, Solubilization, and Types of Dispersions in Surfactant/Water/Oil Systems 12

1.1.1 The Change of the Dissolution State of Nonionic Surfactant with Temperature 12

1.1.2 The Effect of Temperature on the Solubilization of Oil in Aqueous Solutions of Nonionic Surfactants 19
1.1.3 The Effect of Temperature on the Solubilization of Water in Nonaqueous Solutions of Nonionic Surfactants 22
1.1.4 The Change of Emulsion Type with Temperature 24
1.2 The Effect of Temperature and/or the Hydrophilic Chain Length of Surfactant on the Phase Equilibria and the Types of Dispersions of the Ternary System Composed of Water, Cyclohexane, and Nonionic Surfactant 28
1.2.1 The Effect of Temperature on the Phase Equilibria 28
1.2.2 Characteristic Temperature for the Mutual Dissolution of Oil and Water: Hydrophile–Lipophile Balanced Temperature (HLB Temperature) 32
1.2.3 The Effect of Oxyethylene Chain Length of Nonionic Surfactant on the Phase Equilibria 34
1.2.4 The Effect of Temperature on the Dispersion Types 36
1.3 Solution Behavior of Ionic Surfactant + Cosurfactant/Water/Oil Systems 40
1.3.1 The Change of the Dissolution State of Ionic Surfactant + Cosurfactant with the Compositions 41
1.3.2 The Effect of the Types of Counterions, the Types of Surfactants, and Hydrocarbon Chain Length of Surfactants 43

1.3.3 The Effect of Temperature on the HLB of Ionic Surfactant and Cosurfactant Mixture 48
References 50

2 Concepts of HLB, HLB Temperature, and HLB Number 55

2.1 Concepts of Hydrophile–Lipophile Balance (HLB) of Surfactant 56
2.2 HLB Temperature (Hydrophile–Lipophile Balance Temperature or PIT) 58
2.3 HLB Number 68
2.3.1 H/L Number 71
2.3.2 Correlation Between the HLB Number and Other Properties 83
2.3.3 Application of the HLB Method 84
2.4 Hydrophile–Lipophile Balance, HLB, of Ionic Surfactants 88
References 91

3 Factors Affecting the Phase Inversion Temperature (PIT) in an Emulsion 95

3.1 Influence of the Types of Oils on the PIT 96
3.2 Effect of the Oxyethylene Chain Length of Emulsifier on the PIT 99
3.3 The Effect of Phase Volume on the PIT of Emulsions Stabilized with Nonionic Emulsifiers 103
3.4 The Effect of the Hydrocarbon Chain Length of Oils on the PIT of Emulsions 108
3.5 The PIT as a Function of the Composition of Oil Mixtures 108

3.6 The PIT of Emulsions of Emulsifier Mixtures 110
3.7 The Effect of Added Salts, Acid, and Alkali on the PIT of Emulsions 117
3.8 The Effect of Additives in Oil on the PIT 122
References 123

4 Stability of Emulsion 125

4.1 Initial Droplet Diameter and Stability of O/W-Type Emulsions as Functions of Temperature and of PIT (HLB Temperature) of Emulsifiers 126
4.1.1 Emulsification by PIT Method 129
4.1.2 Comparison of Emulsions Prepared by Simple Shaking and Those by the PIT Method 132
4.2 Emulsifier Selection and the Stability of W/O-Type Emulsions as Functions of Temperature and Hydrophilic Chain Length of Emulsifier 136
4.3 The Effect of the Size of Emulsifier and the Distribution of the Oxyethylene Chain Length of Nonionic Emulsifiers on the Stability of Emulsions 144
4.3.1 The Effect of the Size of the Hydrophilic and Lipophilic Moieties on the Stability of Emulsions 145
4.3.2 The Effect of the Distribution of the Hydrophilic Chain Lengths of Emulsifiers on the Stability of Emulsions 148

4.3.3 The Effect of the Substitution of $C_9H_{19}C_6H_4O(CH_2CH_2O)_{8.6}H$ with $C_{12}H_{25} \cdot C_6H_4SO_3Ca_{1/2}$ 155

References 157

5 Liquid Crystals and Emulsions 159

References 168

Index 171

Emulsions and Solubilization

Introduction

Many works have been published on the problems of emulsion (1–9). However, none of them illustrate the specific characteristics of nonionic surfactants of the type polyethylene glycol alkyl(aryl)ethers. These specific properties are essential for their practical use and we found a text with a systematic description of them to be a necessary and useful addition to the literature. This book illustrates the following properties:

1. The effect of temperature on systems containing nonionic surfactants has been studied extensively because the solution properties of the system change drastically and the emulsion type inverts from oil/water (O/W) to water/oil (W/O) with temperature elevation. Nonionic surfactant is hydrophilic at lower temperature and lipophilic at higher temperature.

2. A temperature rise in nonionic surfactant solution corresponds to a shortening of the hydrophilic chain length of the surfactant. Hence, the effect of hydrophilic chain length of nonionic surfactants on various properties can be learned from the studies of one surfactant as a function of temperature.

3. For the control of the hydrophilic–lipophilic balance (HLB) of ionic surfactants, the following products may be used: (a) ionic surfactants soluble in hard water; (b) more lipophilic, oil soluble, cosurfactants; and (c) well-balanced ionic surfactants. Ordinary ionic surfactants are too hydrophilic for such uses, but by control of the HLB, that is, the addition of salt and lipophilic cosurfactant, the solution properties of ionic surfactant may be changed continuously from water soluble to oil soluble.

4. Regardless whether ionic or nonionic surfactants are used, the solubilization is maximum, oil/water interfacial

tension is minimum, and emulsions are most stable when HLBs of surfactants are at respective optimum values. The optimum HLB of surfactant in the system is the key factor in order to reduce the maximum ability of the surfactant solutions.

This book is written to convey these systematic developments.

1.1 DEFINITION OF EMULSIONS

Clayton defines an emulsion as (10):

> An emulsion is a system containing two *liquid* phases, one of which is dispersed as globules in the other. That liquid which is broken up into globules is termed the dispersed phase, whilst the liquid surrounding the globules is known as the continuous phase or dispersing medium. The two liquids, which must be immiscible or nearly so, are frequently referred to as the internal and external phases, respectively.

He further stated that when one of the liquids is water and the other a water-insoluble liquid or "oil," two sets of emulsions are theoretically possible, depending upon whether oil is dispersed in water, O/W, or vice versa, W/O.

Becher's definition is as follows (11):

> An emulsion is a heterogeneous system, consisting of at least one immiscible liquid intimately dispersed in another in the form of droplets, whose diameter, in general, exceeds 0.1 μ. Such systems possess a minimal stability, which may be accentuated by such additives as surface-active agents, finely divided solids, etc.

The IUPAC definition is (12):

> An *emulsion* is a dispersion of droplets of one liquid in another one with which it is incompletely miscible. Emulsions of droplets of an organic liquid (an "oil") in an aqueous solution are indicated by the symbol O/W and emulsions of aqueous droplets in an organic liquid as W/O. In emulsions the droplets often exceed the usual limits for colloids in size.

When the hydrophile–lipophile property of a surfactant balances for a given water + oil, three phases (water, surfactant and oil) coexist. If the amount of oil and water is decreased, two phases exist, the W/D and O/D phases. In such a condition, emulsion type or continuous phase is not as simple as for O/W or W/O only (13). Surfactant phase often occupies a large volume fraction and may be a bi-continuous phase. Emulsion type is usually W/D or O/D type. When a liquid crystalline phase separates instead of the surfactant phase, there exist more emulsion types.

I.2 DEFINITION OF SOLUBILIZATION

According to McBain and Hutchinson (14):

> Solubilization is the name given by J. W. McBain to a particular mode of bringing into solution substances that are otherwise insoluble in a given medium. Solubilization involves the previous presence of a colloidal (organized) solution whose particles (organized) take up and incorporate within or upon themselves the otherwise insoluble material (14).

I.3 DEFINITION OF MICROEMULSION

Prince, former associate of Schulman, said (15):

> When Schulman coined the term micro emulsion, he used as his frame of reference the emulsions with which he and Cockbain had worked (16). These were "fine" emulsions in the droplet size range of 0.5 to 4 microns and could be seen in the light microscope. They scattered white light, i.e., they were opaque like milk and separated on standing. On the other hand, the dispersions he called micro emulsions did not separate and were transparent or translucent (opalescent). This put the diameter of the particle below $^1/_4\ \lambda$, i.e., below 1400 Å. Since these fluid W/O and O/W systems did not exhibit optical, streaming birefringence, Schulman considered the dispersed phase to be in the form of spherical droplets. He measured the size of these droplets by the means available to him at that time, low angle x-ray scattering, light scattering and sedimentation velocity. In 1958, upon seeing electron micrographs of spherical metallic skeletons of the droplets of O/W alkyd emulsions, the diameters of which were in the 75 to 1200 Å range, he coined the term microemulsion to describe these stable dispersions.

Hence, in a strict (narrow) sense, a microemulsion may be defined as a system of water, oil, and amphiphile(s) which is one phase, thermodynamically stable with relatively large swollen micelles (17–23). In a wide sense it may include fairly stable dispersions that look like microemulsions, that is, transparent or translucent (opalescent) (24–27).

Unlike Schulman's initial emphasis (transparency), important features of microemulsions may be their thermodynamic stability and high solvent power (28).

1.4 DEFINITION OF HLB, HLB NUMBER, AND HLB TEMPERATURE (PHASE INVERSION TEMPERATURE = PIT IN EMULSION)

HLB. As Clayton has drawn attention to the concept of a balanced emulsifying agent embodied in a series of patents dating back to 1933 (29), the hydrophile–lipophile balance (HLB) of surfactant in the emulsion is a concept studied and used by many emulsion scientists. The HLB of surfactant in the system certainly changes with temperature, the types of oils, the types and amount of additives in water and in oil, and so on.

HLB number. Based on a large number of experiments, Griffin assigned an HLB number to each surfactant. Due to the definition of Griffin (30), it is apparent that an HLB number is not a function of variables such as temperature, pressure, salt concentration in solution, and types of oils. Therefore, the required HLB of oils is necessary to select emulsifier; it is specific for the nature of the oil, the temperature, other additives in the two phases, and so on.

HLB temperature. In contrast to the HLB number, the HLB temperature (or PIT = phase inversion temperature in emulsion) is a characteristic property of an emulsion (rather than of the surfactant molecule considered in isolation) at which the hydrophile–lipophile property of nonionic surfactant just balances. The effect of additives on the solvent, the effect of mixed emulsifiers or of mixed oils, and so on, are all reflected in the PIT and automatically adjusted for in its determination, and thus, illustrates the change of the HLB of the emulsifier at the interface (31).

It is to be expected that there should be a correlation between the HLB temperature and the HLB number. Such correlations are known. We can determine the HLB number from the HLB temperature of a surfactant with the aid of such data (32,33) and vice versa.

In the case of ionic surfactants, however, the HLB does not change significantly with temperature. But, it is possible to achieve such changes by replacing some ionic surfactant with carefully selected cosurfactant (28).

REFERENCES

1. P. Becher, *Emulsions: Theory and Practice*, 2nd ed., Reinhold, New York (1966).
2. P. Sherman (Ed.), *Emulsion Science*, Academic, New York (1968).
3. K. J. Lissant (Ed.), *Emulsions and Emulsion Technology*, Marcel Dekker, New York, Parts I and II (1974), Part III (1984).
4. S. E. Friberg (Ed.), *Food Emulsions*, Marcel Dekker, New York (1976).
5. P. Becher and M. N. Yudenfreund (Eds.), *Emulsions, Lattices and Dispersions*, Marcel Dekker, New York (1978).
6. P. Becher (Ed.), *Encyclopedia of Emulsion Technology*, Vol. I, Marcel Dekker, New York (1983).
7. L. M. Prince, *Microemulsions: Theory and Practice*, Academic, New York (1977).
8. I. D. Robb (Ed.), *Microemulsions*, Plenum, New York (1982).
9. K. Shinoda (Ed.), *Solvent Properties of Surfactant Solutions*, Marcel Dekker, New York (1967).
10. W. Clayton, *The Theory of Emulsions and Their Technical Treatment*, 4th ed., The Blakiston Co., New York (1943), p. 1.
11. Reference 1, p. 2.

12. *International Union of Pure and Applied Chemistry*, Manual on Colloid and Surface Chemistry, Butterworth, New York (1972).
13. K. Shinoda and H. Saito, *J. Colloid Interface Sci.*, **26**, 70 (1968).
14. M. E. L. McBain and E. Hutchinson, *Solubilization and Related Phenomena*, Academic, New York (1955).
15. Reference 7, p. 18.
16. J. H. Schulman and E. G. Cockbain, *Trans. Faraday Soc.*, **36**, 551 (1940).
17. I. Danielsson and B. Lindman, *Colloids Surfaces*, **3**, 391 (1981).
18. Reference 9, pp. 3–4.
19. G. Gillberg, H. Lehtinen, and S. E. Friberg, *J. Colloid Interface Sci.*, **33**, 40 (1970).
20. P. Ekwall, L. Mandell, and K. Fontell, *J. Colloid Interface Sci.*, **33**, 215 (1970).
21. K. Shinoda and H. Kunieda, *J. Colloid Interface Sci.*, **42**, 382 (1973).
22. K. Shinoda and S. E. Friberg, *Adv. Colloid Interface Sci.*, **4**, 281 (1975).
23. E. Sjöblom and S. E. Friberg, *J. Colloid Interface Sci.*, **67**, 16 (1978).
24. S. E. Friberg, *Colloids Surfaces*, **4**, 201 (1982).
25. W. E. Gerbacia and H. L. Rosano, *J. Colloid Interface Sci.*, **44**, 242 (1973).
26. W. E. Gerbacia, H. L. Rosano, and M. Zajac, *J. Am. Oil Chem. Soc.*, **53**, 101 (1976).
27. M. Podzimek and S. E. Friberg, *J. Dispersion Sci. Technol.*, **1**, 34 (1980).
28. K. Shinoda, H. Kunieda, T. Arai, and H. Saito, *J. Phys. Chem.*, **88**, 5126 (1984).

29. Reference 10, p. 127.
30. W. C. Griffin, *J. Soc. Cosmet. Chem.*, **1**, 311 (1949); **5**, 249 (1954).
31. K. Shinoda and H. Arai, *J. Phys. Chem.*, **68**, 3904 (1964).
32. H. Arai and K. Shinoda, *J. Colloid Interface Sci.*, **25**, 399 (1967).
33. K. Shinoda, Proc. 5th Int. Congr. Surface Active Substances, Barcelona, **3**, 275 (1968); Chapter 5 in Ref. 6, pp. 352–353.

CHAPTER 1

Correlation Between Solution Behavior of Surfactants and Solubilization Microemulsions or Emulsion Types in Surfactant/Water/Oil Systems

This chapter describes the relationships among the dissolution state of surfactant, the solubilization of oil (or water) in aqueous (or nonaqueous) solution of surfactants, the types of emulsions and phase-inversion temperature (PIT) in an emulsion as a function of temperature, and of the hydrophile–lipophile balance (HLB) of surfactants.

The solution of an optimal surfactant combination with a suitable HLB is the most important factor in order to obtain a larger solubilization in micellar systems, ultralow interfacial tension, or emulsions with controlled stability.

Temperature is the most important variable for water/oil (W/O) systems with nonionic surfactant (Sections 1.1 and 1.2) while the system with hydrophilic ionic surfactant + lipophilic cosurfactant shows little or no dependence on temperature (Section 1.3). The HLB of these combinations is instead varied by changing the composition.

1.1 THE EFFECT OF TEMPERATURE AND/OR THE HYDROPHILIC CHAIN LENGTH OF EMULSIFIER ON THE SOLUTION BEHAVIOR OF NONIONIC SURFACTANT, SOLUBILIZATION, AND TYPES OF DISPERSIONS IN SURFACTANT/WATER/OIL SYSTEMS

1.1.1 The Change of the Dissolution State of Nonionic Surfactant with Temperature

Nonionic surfactant added to the two-phase system of water and hydrocarbon preferentially adsorbs at the interface, forming an adsorbed monolayer. The unadsorbed surfactant dissolves either in the water or in the hydrocarbon phase, depending on the temperature of the solution and the hydrophile–lipophile balance of surfactant molecule. A nonionic surfactant with a polar part, an oxyethylene chain, usually dissolves in the oil phase at high temperature, whereas it dissolves in water at low temperature. Over a narrow medium temperature range in which the solubility of the surfactant is extremely small in water and low in the oil phase, a third phase, an isotropic liquid is separated.

The convex or concave curvature of the adsorbed monolayer against water (or oil) seems closely related to the dissolution state of nonionic agent and to the phase diagrams in the three-component system composed of water, hydrocarbon, and nonionic agent (1). The hydration forces between the hydrophilic moiety of surfactant and water are stronger at lower temperatures, and the adsorbed monolayer may have a convex curvature towards water. Since a stronger interaction means greater affinity and zero or small interfacial energy, the consequent increase in interface does not result in a large free energy increase of the system. This concept nicely correlates with O/W-type emulsions at low temperature (2). On the other hand, the decrease of oil/surfactant interface contributes more efficiently to the free energy decrease of the system at the lower temperature. Thus, the convex curvature towards water may be thermodynamically preferable. The dissolution of a nonionic surfactant in an aqueous phase as micelles below the cloud point is regarded as a similar phenomenon.

This treatment essentially emphasizes the importance of the bending surface energy for this type of dispersion; an important concept (3). As a matter of fact, the microemulsion literature for a long time considered it the only thermodynamic quantity of importance (4). Although later contributions (5–11) have shown this thought to be an oversimplification, the surface free energy terms are extremely important for the stability of a microemulsion or micellar system. Murphy (12) has given a thorough and competent discussion of the stretching, bending, and torsional components of the surface free energy of a liquid/liquid interface.

Although a micellar solution containing hydrocarbon within the limit of solubilization is transparent* and infinitely stable in contrast to an emulsion, there are similarities between the structure of micelles in a solution with large solubilization of hydrocarbon and that of emulsion droplets. A surface zone and a central core have distinctly different properties for both kinds of particles. Addition of hydrocarbon beyond the solubilization limit results in an O/W-type emulsion, with excess oil being dispersed as droplets. The nonionic surfactant adsorbed on the interface has a convex curvature toward water at that temperature.

Figure 1.1A illustrates schematically the state of the solution composed of water, nonionic surfactant, and oil below the cloud point. Figure 1.1, as a whole, is designed schematically to show the change in the types of dispersion of the three-component system, water/hydrocarbon/nonionic agent, with temperature. If the temperature of an aqueous solution of a nonionic surfactant is raised, the hydration between water and the hydrophilic moiety gradually decreases and the hydrophile–lipophile balance changes towards more lipophilic character. The convex curvature of the adsorbed monolayer towards water will gradually change to a concave curvature. At some temperature where the curvature of the adsorbed monolayer is on an average zero, the nonionic sur-

*Large solubilization may give solutions that are transparent or translucent in transmitted light, but which may appear turbid in the perpendicular direction.

factant aggregates infinitely with a large amount of water and a phase separation occurs. This phenomenon in an aqueous solution of a common nonionic surfactant is called clouding, and the temperature is, of course, the cloud point.

With an aliphatic hydrocarbon present the cloud point temperature is increased as shown in Section 1.1.2 and as the surfactant molecules associate to aggregates with more complex structures, the solution splits into two phases called the surfactant phase and the aqueous phase. The surfactant phase contains large amounts of water, hydrocarbon, and surfactant concurrently; (13) the total system changes to three phases as shown in Fig. 1.1B and 1.1C.

The phase equilibria with high water concentrations form a fairly intricate pattern in the presence of an aliphatic hydrocarbon. Figures 1.2 and 1.3 schematically show the consecutive developments with temperature from the micellar solution to the formation of a separate surfactant phase.

The features of a water, aliphatic hydrocarbon, nonionic surfactant system at the temperature range in which the cloud point is enhanced by the presence of the hydrocarbon are illustrated in Fig. 1.2. At the cloud point the surfactant/water micellar solution is separated into two solutions of which one contains extremely small amounts of surfactant which instead is concentrated in the second solution. When this happens (A → B in Fig. 1.2) an "internal" two-phase area is developed in the micellar area. Hence, a sufficiently high concentration of hydrocarbon means an unbroken solubility area to 100% water, for example, the cloud point phenomenon is overcome.

The continued development of the formation of the surfactant phase displays an interesting effect. With further increase of temperature the highly divided micellar area in Fig. 1.2C separates into two areas according to Fig. 1.3A.

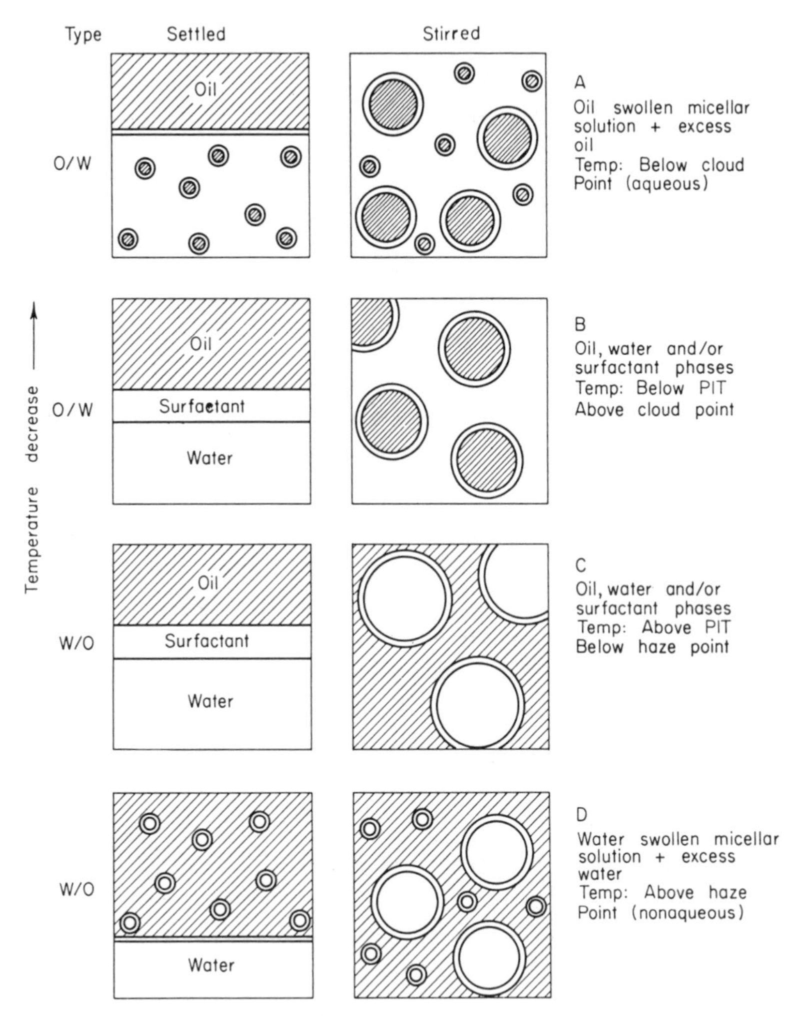

Figure 1.1 Schematic diagrams of the change of the dispersion state with temperature in a solution composed of water, hydrocarbon, and nonionic surfactant. Small circles indicate swollen micelles. Large circles indicate emulsion droplets. [Reproduced by permission of Academic, *J. Colloid Interface Sci.* **24**, 4 (1967).]

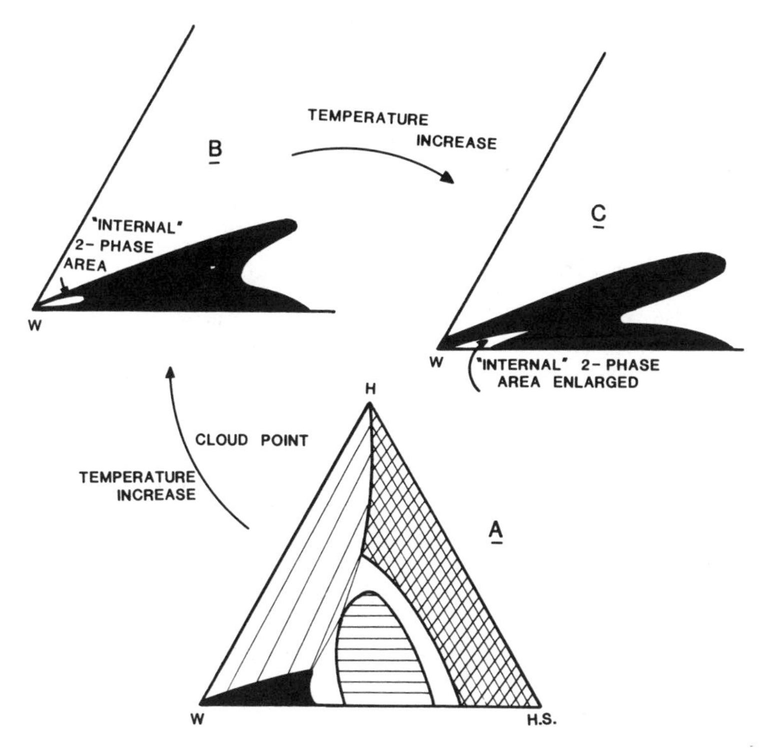

Figure 1.2 When the temperature is increased passing the cloud point (Diagram A → B) the infinite solubility of water in the aqueous micellar phase disappears. With a certain amount of hydrocarbon present (B) the infinite solubility of water is restored; hence, addition of hydrocarbon raises the cloud point. Further increase of the temperature (B → C) leads to an enlargement of this "internal" two-phase area and an enlarged "dent" from the direction of the liquid crystal.

The upper area (P, Fig. 1.3A) may be called an O/W microemulsion. With further increase of the temperature (Fig. 1.3A to 1.3B) this microemulsion phase will separate from the aqueous corner and a three-phase area is formed.

A surfactant phase is now in equilibrium with an aqueous solution with an extremely low concentration of the surfac-

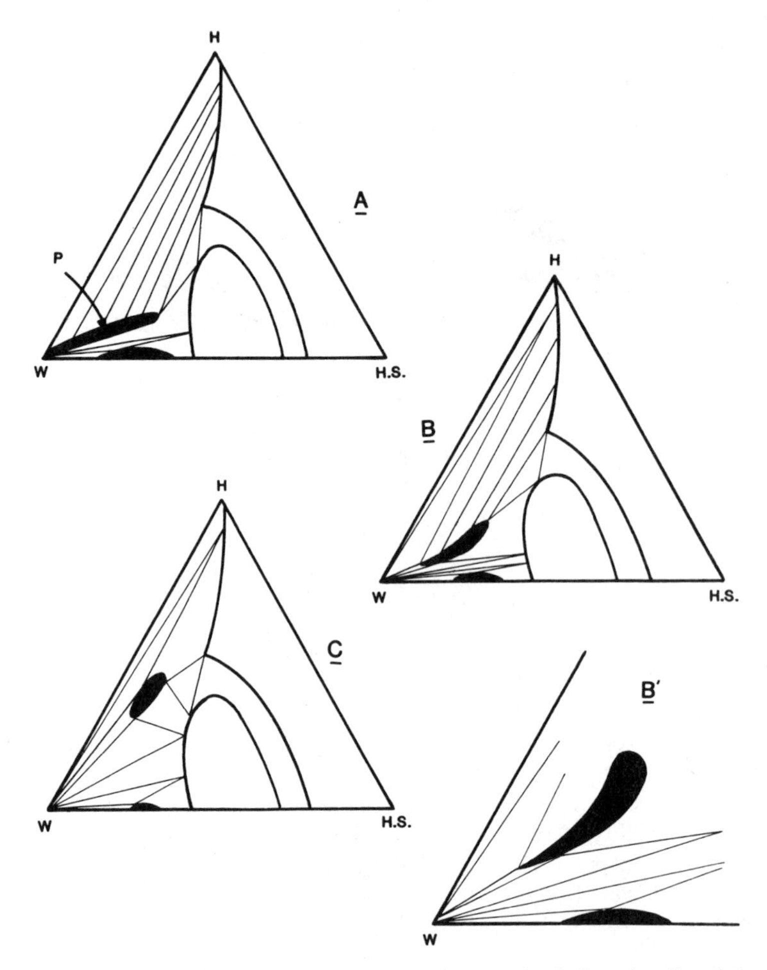

Figure 1.3 With further increased temperature both the "dent" and the "internal" two-phase area in Fig. 1.2C will be enlarged and two separate areas occur as in the A part of this figure. The upper part (P) is not in equilibrium with the lower part except at extremely low concentrations of hydrocarbon and surfactant. The equilibrium is instead with the lamellar liquid crystal. Further increase of the temperature leads to a disconnection at the water corner (B) and the formation of the separate surfactant phase. This phase region will with further increase of the temperature move towards lower surfactant concentrations and high water *and* hydrocarbon content (C).

tant and a hydrocarbon phase with 1 or 2% of surfactant and some solubilized water. The separation of the surfactant phase may be called the second cloud point in the system and it is obvious that a careful reduction of the temperature from state B in Fig. 1.3 toward state A will mean a strong reduction of the interfacial tension between the aqueous phase and the surfactant phase. It should be observed that this means obtaining ultralow interfacial tension with extremely small concentrations of the surfactant; fractions of a percent.

The state in Fig. 1.3C means a microemulsion, the surfactant phase, with a minimum of surfactant to bring large amounts of hydrocarbon *and* water concurrently into a microemulsion. It should be realized that such microemulsions are extremely temperature sensitive, there are different means of improving the temperature range of stability (14,15).

1.1.2 The Effect of Temperature on the Solubilization of Oil in Aqueous Solutions of Nonionic Surfactants

The preceding model is confirmed by the temperature dependence of the solubilization of hydrocarbon in aqueous solution of nonionic surfactants presented in Fig. 1.4 (16).

The filled circles in Fig. 1.4 are the cloud points of surfactant solution as a function of the amount of added hydrocarbon; the open circles illustrate the solubilization limit of oil. The solubilized region between two curves is a homogeneous phase in which hydrophilic micelles with solubilized oil are dispersed in water. The region is transparent when the micelles are small. But, the turbidity of the solution continuously increases proportionally with the increase in solubilizate, because the turbidity (τ) of micellar solution depends

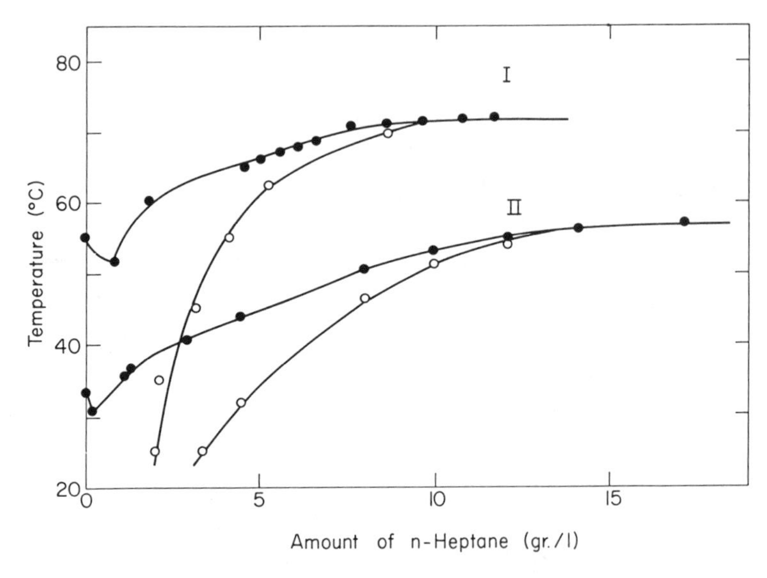

Figure 1.4 The effect of temperature on the solubilization of *n*-heptane in 1% aqueous solution of I, polyoxyethylene(9.2)nonylphenylether and II, polyoxyethylene(9.0)dodecylphenylether. Filled circle means cloud point and open circle means solubilization limit. [Reproduced by permission of Academic, *J. Colloid Interface Sci.*, **24**, 4 (1967).]

on the size (V), number (n), and difference in refractive index ($\Delta\rho$) of oil containing micelle against medium (17).

$$\tau = \frac{32\pi^3 \rho_0^2 (\Delta\rho)^2 n V^2}{3\lambda^4} \tag{1}$$

where λ is the wavelength of light.

Excess oil separates or is emulsified as an O/W-type emulsion at the temperature below the solubilization curve, that is, the system is composed of micellar solution and an oil phase (Fig. 1.2). On the other hand, the surfactant phase separates above the cloud point curve provided the surfactant concentration is sufficiently small. For higher concentrations of surfactant the phase equilibria are complex (Fig. 1.3B and 1.3C).

In the case where the solubilizate is completely miscible with only the surfactant phase, two phases exist above the cloud point, but in the case where the solubilizate does not dissolve completely in the surfactant phase, three phases coexist at higher solubilizate concentration (Fig. 1.2). The cloud point of surfactant solution containing hydrocarbon

Figure 1.5 Schematic illustration of the structure of surfactant phase. Black and white stripes represent the bound water and oil layers, respectively. There exist surfactant monolayers between bound water and oil layers. [Reproduced by permission of Steinkoff Verlag-Darmstadt, *Progr. Colloid Polymer Sci.*, **68**, 1 (1983).]

TABLE 1.1 The Cloud Points of 1% Aqueous Solutions of Polyoxyethylene(9.2) Nonylphenyl Ether Saturated with Various Oils

Types of oil	Cloud point (°C)	Types of oil	Cloud point (°C)
No oil	56	Liquid paraffin	80.4
n-$C_{16}H_{34}$	80	C_6H_{12}(cyclo)	54
n-$C_{12}H_{26}$	79.3	CCl_2CCl_2	31
n-$C_{10}H_{22}$	79	$C_2H_5C_6H_5$	30.5
n-C_7H_{16}	71.5	C_6H_6	Below 0

corresponds to a temperature above which the curvature of the adsorbed monolayer changes from convex to flat on the average towards water. Figure 1.5 schematically illustrates the structure of the surfactant phase.

The cloud points of solutions saturated with various hydrocarbons differ widely with the kinds of hydrocarbons as shown in Table 1.1 (17).

1.1.3 The Effect of Temperature on the Solubilization of Water in Nonaqueous Solutions of Nonionic Surfactants

Similar phase diagrams are observed in nonaqueous solutions of nonionic surfactants. A typical curve is shown in Fig. 1.6 (18). In this case in contrast with an aqueous solution, water is solubilized in the oleophilic micelle dispersed in the hydrocarbon. Open circles illustrate the solubilization limit, whereas filled circles correspond to the cloud point* in non-

*Below this temperature the nonaqueous surfactant solution becomes *hazy* rather than cloudy if no water is present (*18*). The visual difference between cloud points in aqueous and nonaqueous solutions is fairly distinct.

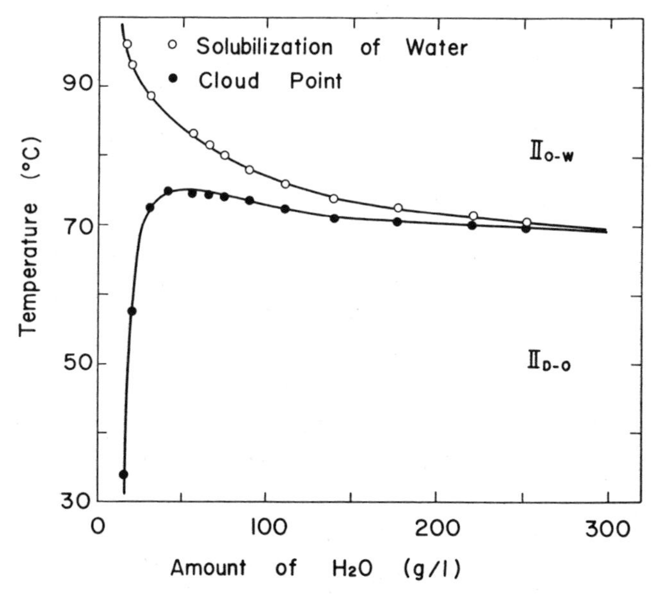

Figure 1.6 The effect of temperature on the solubilization of water in 10% polyoxyethylene(9.6)nonylphenylether/90% cyclohexane solution. Open circles mean solubilization limit and filled circles mean cloud (haze) point. [Reproduced by permission of Academic, *J. Colloid Interface Sci.*, **24**, 4 (1967).]

aqueous surfactant solution as a function of the amount of added water. The cloud point in a nonaqueous surfactant solution is the temperature below which the surfactant phase containing water separates from the nonaqueous solution. The solubilized region between the cloud point curve and the solubilization curve is one phase in which the oleophilic micelles with solubilized water disperse in hydrocarbons.

Excess water disperses as a W/O-type emulsion above the solubilization curve, that is, the adsorbed monolayer of surfactant has concave curvature towards water and convex curvature towards hydrocarbon. Figure 1.1D schematically illustrates the state of the solution above the cloud point in nonaqueous surfactant solution. This concave curvature towards water may decrease with the temperature depression and become an average zero at the cloud point temperature. The cloud point in nonaqueous solution is usually 5–15°C higher than the cloud point in aqueous solution (18). The difference in the two cloud points is small when the surfactant concentration is high (>5 wt%) or the surfactant is pure.

These changes are reflections on the continuation of the series aqueous micellar solution/O/W microemulsion/surfactant phase in Figs. 1.2 and 1.3. The consecutive developments are displayed in Fig. 1.7. At temperatures in excess of the HLB temperature the area for the surfactant phase moves towards the hydrocarbon/surfactant solution with solubilized water (Fig. 1.7A). At a certain temperature the two areas coalesce with each other and the characteristic *solubility maximum* of water in Fig. 1.7B appears. One may now speak of an W/O microemulsion or an inverse micellar solution.

With further elevation of temperature the maximum hydrocarbon content of the liquid crystal is reduced and hence, the surfactant/hydrocarbon ratio of *maximum* water solubilization is increased, Fig. 1.7B, 1.7C, and 1.7D. Finally, the liquid crystalline phase disappears and the entire system displays the simple features of Fig. 1.7E encompassing the hydrocarbon/surfactant solution with solubilized water in equilibrium with water.

1.1.4 The Change of Emulsion Type with Temperature

The emulsion stabilized with a nonionic surfactant is a W/O type at high temperature but an O/W type at low tempera-

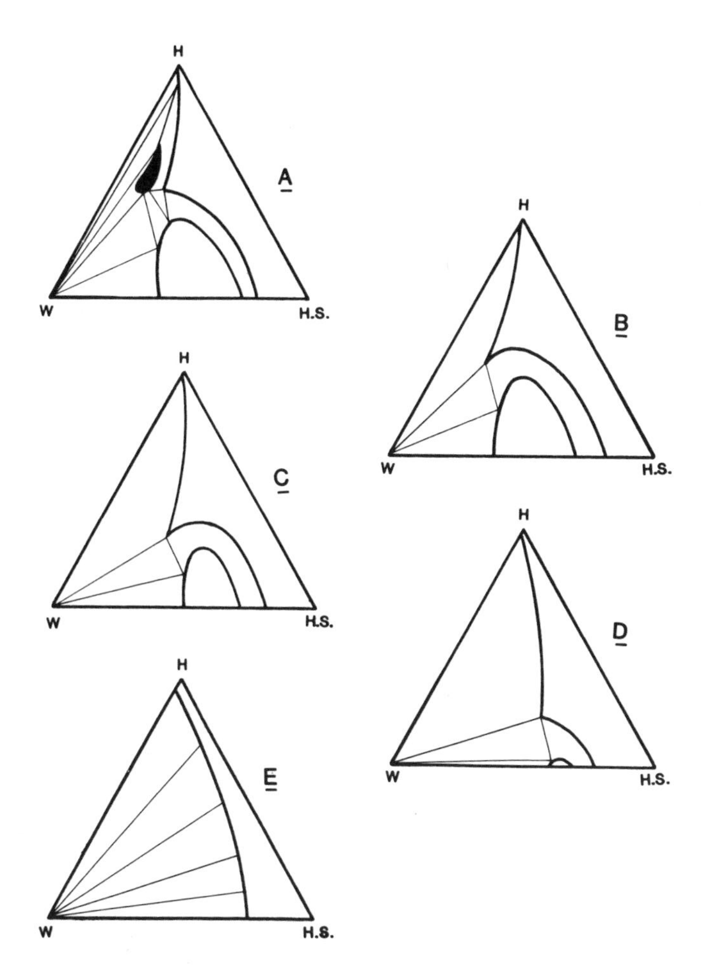

Figure 1.7 The area for the surfactant phase (black, A) coalesces with the hydrocarbon/emulsifier solution forming a W/O microemulsion area (B). With further increase of the temperature the hydrocarbon/surfactant ratio for the maximum water content is reduced (B → C → D) and with it the liquid crystalline area. Finally the latter one disappears entirely and only a two-phase area (E) remains.

ture, and a phase inversion in the emulsion takes place at some medium temperature. The existence of a phase inversion temperature (PIT) indicates appreciable changes of the hydrophilic–lipophilic balance (HLB) of a nonionic surfactant with temperature. This influence is strong; a change of the HLB of the surfactant or the type of emulsion is obtained with altered temperature; in fact, the effect of phase volume on the phase inversion may be smaller than that of temperature. This reasoning is clearly supported by the PIT versus the phase volume relation. A typical curve is shown in Fig. 1.8 (19).

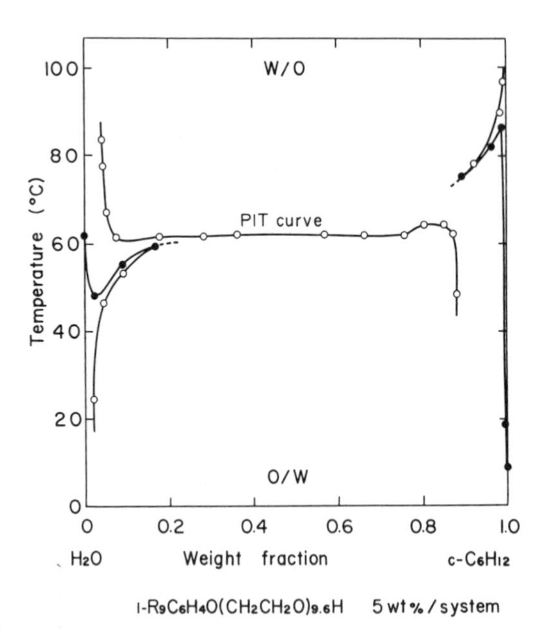

Figure 1.8 The effect of phase volume on the phase inversion temperature (PIT) in a cyclohexane/water system emulsified with 5 wt% of polyoxyethylene(9.6)nonylphenylether. [Reproduced by permission of Academic, *J. Colloid Interface Sci.*, **24**, 4 (1967).]

The PIT stays almost constant over a wide volume fraction range, indicating the strong type-determining tendency of the adsorbed monolayer of nonionic surfactant. The emulsion is an O/W type below the PIT curve and a W/O type above the PIT curve. The reproducibility of the PIT also indicates the total effect of the temperature on the emulsion drop curvature. It is important to realize the influence of the surfactant on the emulsion drop curvature as indirect. The bending energy (12) is not important for the curvature of an emulsion droplet with a radius of several microns; the influ-

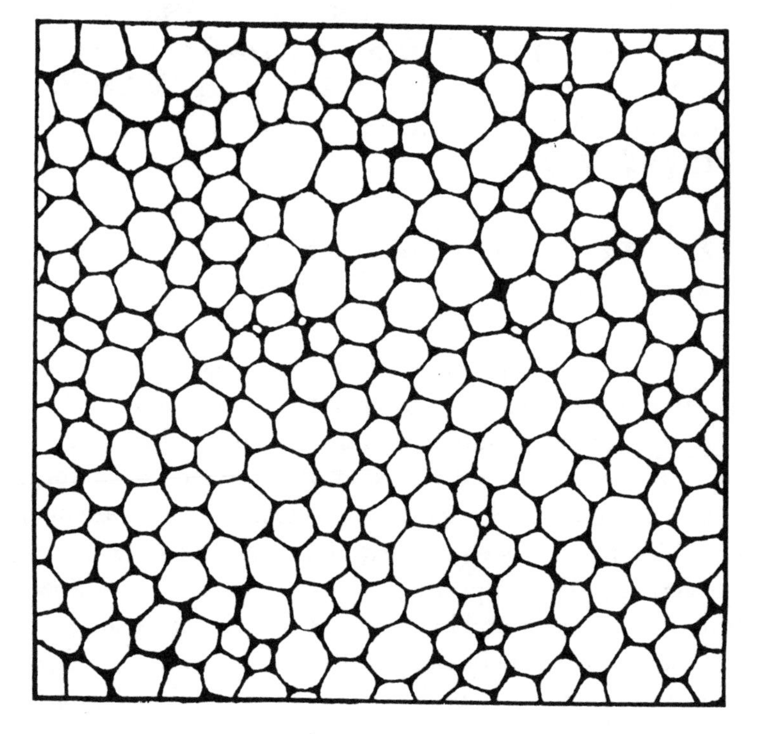

Figure 1.9 Schematic representation of a close-packed emulsion in which the volume fraction of dispersed phase is large.

ence arises from the differences in stability of W/O and O/W systems during formation (20). An O/W system is obtained because the W/O system is highly unstable.

The volume fraction of the continuous phase at both ends of the phase inversion curve is so small that the emulsion droplets become close packed. The system behaves like a gel. The schematic representation of such an emulsion is shown in Fig. 1.9.

1.2 THE EFFECT OF TEMPERATURE AND/OR THE HYDROPHILIC CHAIN LENGTH OF SURFACTANT ON THE PHASE EQUILIBRIA AND THE TYPES OF DISPERSIONS OF THE TERNARY SYSTEM COMPOSED OF WATER, CYCLOHEXANE, AND NONIONIC SURFACTANT

1.2.1 The Effect of Temperature on the Phase Equilibria

The knowledge of phase equilibria in ternary systems composed of water, hydrocarbon, and nonionic surfactant is important in order to understand the dissolution behavior, the type of dispersion in an emulsion, and the mutual solubility of water and oil by the action of surfactants. There are many variables in such systems such as the types of oils, the kinds of surfactants, the composition of components, and temperature. Among saturated hydrocarbons, cyclohexane may well represent typical behavior in these systems. Although the composition of water versus oil has to be explored over the entire volume fraction range, the concentration of the surfactant may be fixed at 1–10% from a practical viewpoint

(1). However, complete phase diagrams of three-component systems are necessary in order to understand the delicate phase equilibria at higher surfactant concentrations (11,21–24) (Figs. 1.2, 1.3, and 1.7).

Although little attention has been paid to the effect of temperature in the field of emulsions until 1963, it has been proven that the studies of the effect of temperature on the solubilization, emulsion type, and solution behavior of nonionic surfactant was utterly necessary to clarify various phenomena in these fields (1,2,14,25). Because the change of temperature in a solution of nonionic surfactant parallels the change of the HLB or oxyethylene chain length of nonionic surfactant, the studies of the effect of temperature affords information also on these factors in a simple and efficient manner. The change in either hydrophilic or lipophilic chain length of the surfactant shifts the phase equilibria and dispersion types either to higher or lower temperatures, but the pattern is similar.

The study of these systems is useful in understanding the mutual relations between: (a) solubilization of oil in aqueous surfactant solutions; (b) solubilization of water in nonaqueous surfactant solutions; (c) the types, inversion, and stability of emulsions; and (d) practical applications, such as washing, dry cleaning, and emulsification.

Hence, a typical surfactant such as polyoxyethylene-(9.7)nonylphenylether represents the general behavior in these ternary systems, and phase equilibria and dispersion types of water/cyclohexane systems containing 3 and 7 wt% per system of $i\text{-}C_9H_{19}C_6H_4(CH_2CH_2O)_{9.7}OH$ as a function of temperature has been used as a model system (1). The phase diagram of a water/cyclohexane systems containing 7 wt% polyoxyethylene(9.7)nonylphenylether is shown in Fig. 1.10.

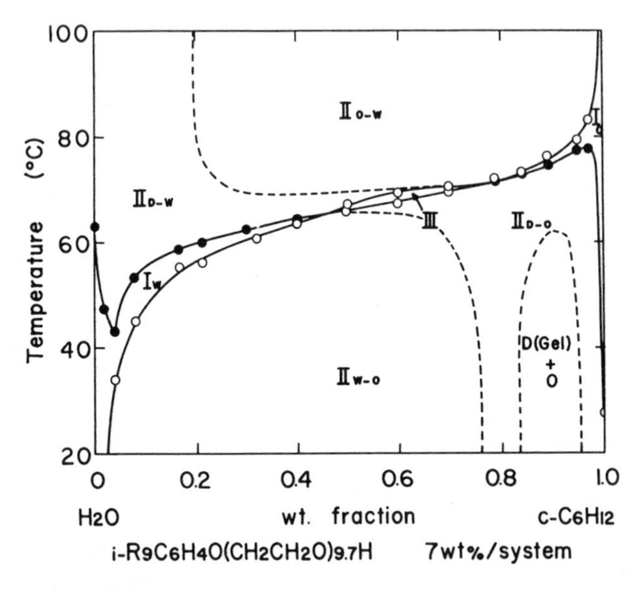

Figure 1.10 The phase diagram of a water/cyclohexane system containing 7 wt% of polyoxyethylene(9.7)nonylphenylether as a function of temperature. [Reproduced by permission of Academic, *J. Colloid Interface Sci.*, **26**, 70 (1968).]

Realm I_W is the micellar solution with solubilized oil. At a relatively low temperature the solubilization of cyclohexane in an aqueous surfactant solution increases markedly close to the cloud point; a surfactant phase and a water phase separate at a temperature level above the cloud point. A large amount of water and cyclohexane dissolves in the surfactant phase, and the two phases (water and surfactant phases) coexist above the cloud point curve, realm $II_{D/W}$ (compare Fig. 1.3). If the amount of oil in the system is increased at this temperature, an oil phase appears. The central realm indicated by III represents a three-phase region composed of water, surfactant, and oil phases. As the volume fraction of the surfactant phase is large (about 80% in a 7 wt% solution), the water or oil phase will disappear, owing to a small change of composition or temperature (Fig. 1.3C). Hence, the three-phase realm is narrow. It becomes larger in more dilute solution. To the right-hand side of a three-phase realm there is $II_{D/O}$. The area $II_{D/O}$ is a two-phase region consisting of surfactant and oil phases. Since, with increasing temperature, the solubility of water in a surfactant phase decreases, and that of oil increases, the result is an increase in volume of the water phase and a decrease of surfactant plus oil phase (compare Figs. 1.3C and 1.7).

The right-hand side of Fig. 1.10 corresponds to a nonaqueous solution of a nonionic surfactant containing a small amount of water. Realm I_O is a micellar solution of cyclohexane with solubilized water. The solubilization curve of water in cyclohexane is observed at a relatively high temperature. Solubilization of water increases with a decrease in temperature, particularly near the cloud point in a nonaqueous surfactant solution, but a surfactant phase separates from cyclohexane below the cloud point (Fig. 1.10). Above the I_W, III, and I_O regions two phases exist. As the solubility of the surfactant in water is very small in these regions (or at this temperature range), the aqueous phase is nearly pure water. Accordingly, practically all surfactant is in the oil phase (Fig. 1.7B and 1.7C). The concentration of surfactant in the oil phase increases with the change of composition from the right-hand side to the left-hand side; a tendency that is especially strong around the phase inversion temperature. The right-hand side of the dotted line indicates the two-phase solution consisting of water and oil phases and the left-hand side of the dotted line indicates the two-phase solution consisting of water and surfactant phases.

Similarly, the two phases coexist between the I_W, III, and I_O regions. The surfactant dissolves in water in these regions, but does not dissolve well in oil. At the left-hand side of the two-phase region, excess oil separates from the oil-swollen micellar solution, realm $II_{W/O}$. However, the relative concentration of the surfactant in the water phase increases with the change of composition from left to right, and the surfactant aggregates reach infinite dimensions and a liquid crystal is formed. Hence, the two phases on the right-hand side of the dotted line consist of liquid crystal and oil phases.

1.2.2 Characteristic Temperature for the Mutual Dissolution of Oil and Water: Hydrophile–Lipophile Balanced Temperature (HLB Temperature)

The important feature of solutions of a nonionic surfactant is the notable increase in the solubility of oil in an aqueous surfactant solution at the cloud point in water and of water in a nonaqueous surfactant solution at the cloud point in hydrocarbon. This tendency is also shown in the three-phase realm by the fact that a large amount of oil and water dissolves into the surfactant phase.

Although the aqueous micellar solution (I_W) with solubilized oil, the surfactant phase in realm III, and the hydrocarbon micellar solution (I_O) with solubilized water are three separated realms in the phase diagram, the changes in composition and structure of these phases with temperature are continuous as seen in Fig. 1.11. Realm I_W is a hydrophilic micellar solution, the surfactant phase is deemed to be a micellar solution of an average zero curvature structure, whereas realm I_O is an oleophilic micellar solution. The curvature of surfactant monolayer against oil (or water) seems to change continuously with temperature at constant surfactant concentration. The very small interfacial tension be-

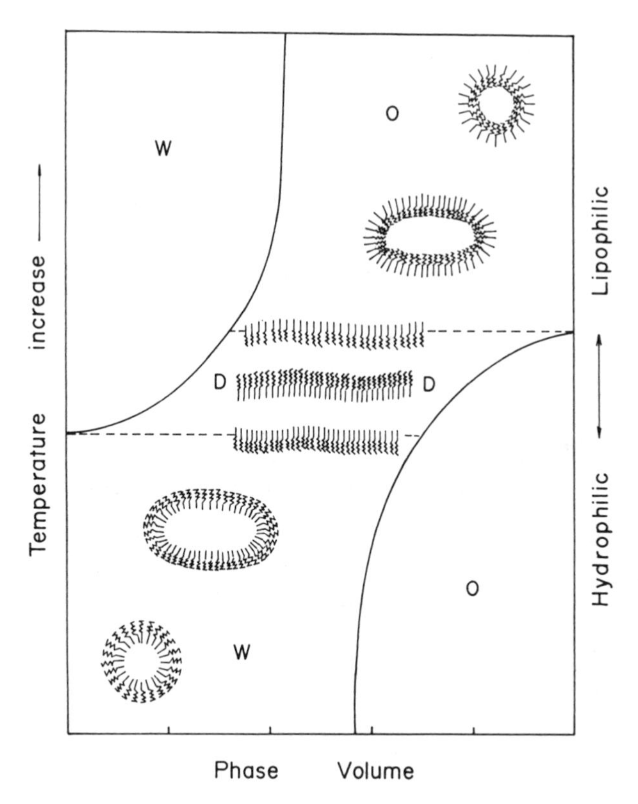

Figure 1.11 Schematic illustration of the change of solution behavior of surfactant with the hydrophile–lipophile balance in a water/surfactant/hydrocarbon system. The change of relative solubility of water and hydrocarbon in a surfactant phase with the change of the HLB of surfactant.

tween surfactant phase and water phase (smaller than 10^{-1} dyne/cm) increases with a temperature decrease (26). This finding suggests that the adsorbed surfactant monolayer at the oil/water interface has a tendency to concavity towards oil at lower temperatures, an average zero curvature at medium temperatures, and to a convexity towards oil at higher temperatures. Although there is no solubilization due to the definition at the temperature of realm III, the solubility of oil in surfactant phases, or that of water in surfactant + oil phases, is high, and detergent action is excellent at this temperature.

The change of the aggregation number of micelles, the solubilizing power, the curvature of surfactant aggregates, and the change of volume fractions of water, oil, and surfactant phases as a function of temperature around the three-phase realm is schematically illustrated in Fig. 1.11.

1.2.3 The Effect of Oxyethylene Chain Length of Nonionic Surfactant on the Phase Equilibria

If the oxyethylene chain length of nonionics is made longer or shorter, a similar phase diagram is obtained and shifted to higher or lower temperature as shown in Fig. 1.12 (19). If the temperature of the system is raised, the interaction between water and the hydrophilic moiety of the surfactant decreases. Thus, the effect of a temperature increase and the decrease in the oxyethylene chain length (hydrophilic property) of the surfactant molecule, may be similar. This reasoning is confirmed by the phase diagram of nonionic surfactants in H_2O plus c-C_6H_{12} system as a function of the average oxyethylene chain length of surfactant and shown in Fig. 1.13 (27). Here the oxyethylene chain length of surfactant in the ordinate decreases instead of temperature.

By varying amphiphiles, additives, hydrocarbons, and compositions, Winsor (28,29) was able to define the limits of completely solubilized systems and the nature of the equilibria of solubilized phases with other phases. He defined equilibria as type I (solubilized phase in equilibrium with dilute hydrocarbon), type II (solubilized phase in equilibrium with

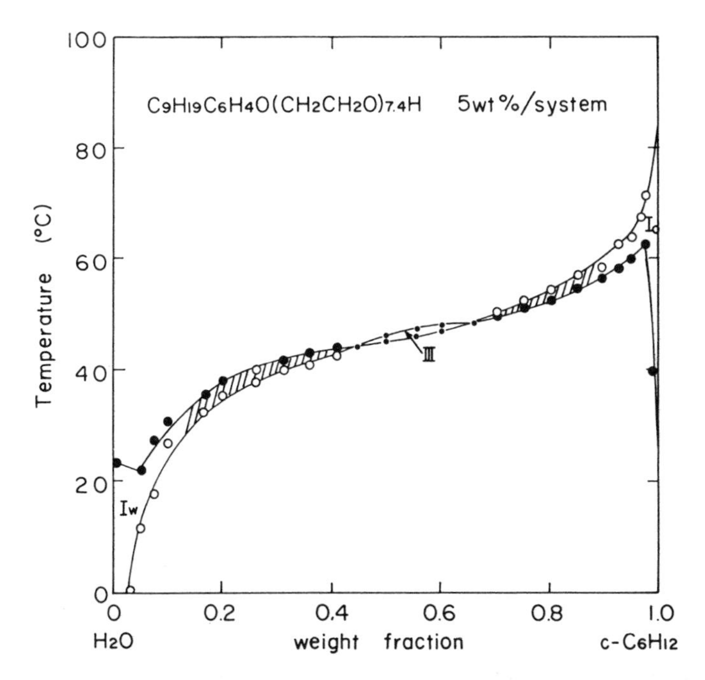

Figure 1.12 The effect of temperature on the phase diagram of a water/cyclohexane system containing 5 wt% of $C_9H_{19}C_6H_4O(CH_2CH_2O)_{7.4}H$.

dilute aqueous phase), type III (solubilized with dilute hydrocarbon and dilute aqueous phases), and type IV (solubilized phase only). Furthermore, type IV systems were subdivided into various isotropic sol and birefringent liquid crystalline phases. Gradual changes in composition led to conversion of one system into another. Realms I_W and I_O in the present study correspond to type IV, realm III to type III, realm $II_{W/O}$ to type I, and realm $II_{O/W}$ to type II of Winsor's classification. Similar solubilized systems have been studied by Schulman and co-workers (30–32) who aptly names these systems microemulsions. The surfactant phase and I_W and I_O phases close to realm III correspond to

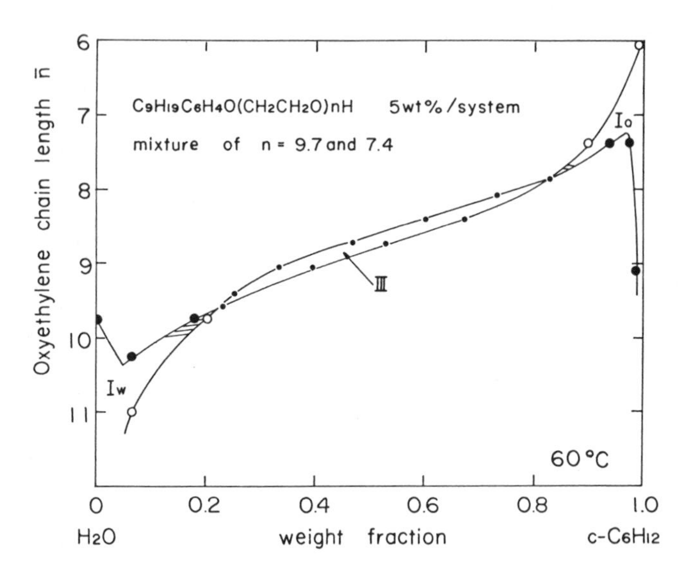

Figure 1.13 The effect of the average oxyethylene chain length of nonionics on the phase diagram of a water/cyclohexane system containing 5 wt% of the mixture of $C_9H_{19}C_6H_4O(CH_2CH_2O)_{7.4}H$ and $C_9H_{19}C_6H_4O(CH_2CH_2O)_{9.7}H$. The temperature increase in Fig. 1.12 and the decrease of oxyethylene chain exhibit practically the same effect. [Reproduced by permission of Academic, *J. Colloid Interface Sci.*, **42**, 381 (1973).]

Schulman's so-called microemulsions. The remarkable difference with Schulman's system is that the concentration of surfactant necessary to obtain a microemulsion is so much smaller (5 wt%) in the present system (1).

1.2.4 The Effect of Temperature on the Dispersion Types

Excess water separates from a nonaqueous micellar solution at a high temperature. The dispersion type of this two-phase solution is a W/O-type emulsion over a wide volume fraction range as shown in Fig. 1.14 (1).

In the region where the volume fraction of oil is smaller than 0.2, the concentration of the nonionic surfactant will be fairly high in the nonaqueous phase and the nonaqueous phase is regarded as a surfactant phase in which hydrocarbon is dissolved. The solution is viscous at this volume fraction.

If the volume fraction of oil further decreases, the water phase (which occupies a very large volume fraction) finally becomes a continuous phase, that is, phase inversion does occur from a W/D(O) to a D(O)/W type.

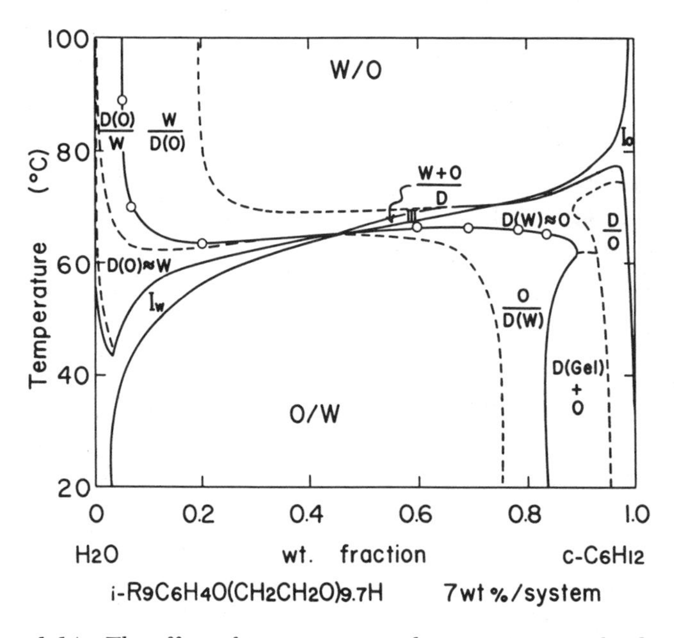

Figure 1.14 The effect of temperature and composition on the dispersion types of water/cyclohexane/polyoxyethylene(9.7)nonylphenylether (7 wt%). [Reproduced by permission of Academic, *J. Colloid Interface Sci.*, **26**, 70 (1968).]

At the medium temperature three phases coexist, so that water, oil, and surfactant phases are more clearly distinguished in realm III. The type of dispersion is the (W + O)/D type; the oil phase disappears on the left-hand side of realm III owing to the decrease of the volume fraction of hydrocarbon, so that the type of dispersion is either a W/D or a D/W type above or below the phase inversion temperature. A region exists between the phase inversion temperature and the cloud point curve in which water and surfactant phases are both continuous (W $\approx$ D) (compare Figs. 1.3A and 1.3B).

On the other hand, the water phase disappears on the right-hand side of realm III, owing to the decrease of the volume fraction of water. The type of dispersion is either a D/O or an O/D type, above or below the phase, respectively, which forms gradually with a temperature decrease or with an increase in the hydrophilic property of the surfactant as shown by the lower dotted line (compare Fig. 1.3B to Fig. 1.3A). The dispersion in the $II_{D/O}$ realm is not always a D/O type above the phase inversion temperature, but both phases are continuous in one part. The type is clearly a D/O type in the region where the volume of oil phase is large, as shown in realm D/O in Fig. 1.14.

At low temperatures, the surfactant dissolves in water and some hydrocarbon is solubilized in aqueous micellar solution. Beyond the solubilization limit excess hydrocarbon disperses as an O/W-type emulsion. Because of the change of composition from the left-hand side to the right-hand side, the concentration of surfactant in the water phase increases, since the amount of surfactant in the system is fixed. Finally, a liquid crystal and oil phase coexist in realm D (gel) + O. If the volume fraction of water further decreases, the stiff gel becomes a sol and a two-phase solution consisting of oil and surfactant phases is obtained. Similar phase equilibria and dispersion types were obtained in the cyclohexane/water system containing 3 wt% of polyoxyethylene(9.7)nonylphenylether per system and are shown in Figs. 1.15 and 1.16.

It can be concluded from Figs. 1.15 and 1.16 that the oil phase is a continuous medium at high temperature (or when an emulsifier is lyophilic), a surfactant phase is a continuous medium at medium temperature close to the phase inversion temperature (HLB temperature or when the hydrophile–lipophile property of surfactant just balances), and a water phase is a continuous medium at low temperature (or when a surfactant is hydrophilic) in systems composed of water, oil, and surfactant.

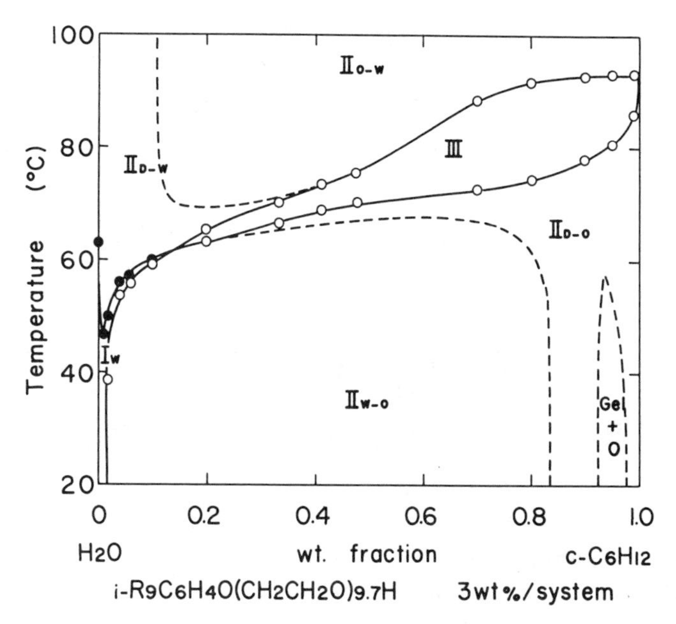

Figure 1.15 The phase diagram of a water/cyclohexane system containing 3 wt% of $C_9H_{19}C_6H_4O(CH_2CH_2O)_{9.7}H$ as a function of temperature. [Reproduced by permission of Academic, *J. Colloid Interface Sci.*, 26, 70 (1968).]

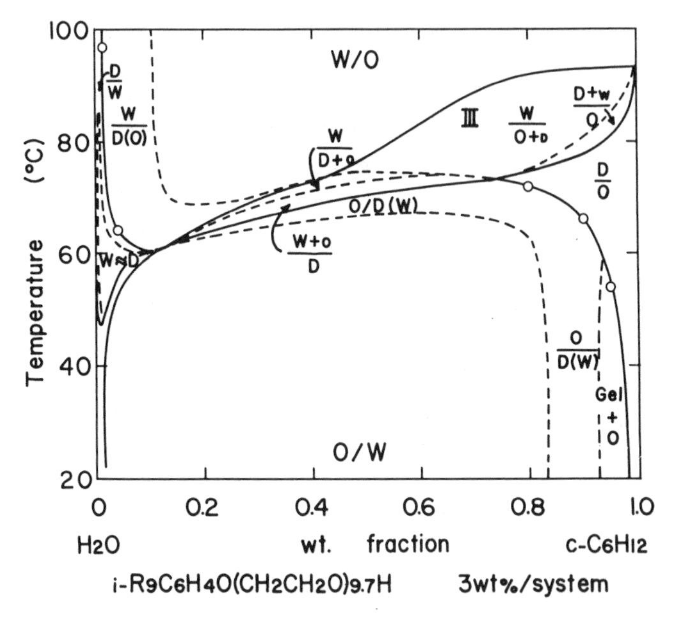

Figure 1.16 The effect of temperature and composition on the dispersion types of a water/cyclohexane system containing 3 wt% of $C_9H_{19}C_6H_4O(CH_2CH_2O)_{9.7}H$. [Reproduced by permission of Academic, *J. Colloid Interface Sci.*, 26, 70 (1968).]

1.3 SOLUTION BEHAVIOR OF IONIC SURFACTANT + COSURFACTANT/WATER/OIL SYSTEMS

In order to attain a large solubilization, a required emulsion type, or to obtain a more stable emulsion, the selection of an optimum hydrophilic chain length of surfactant (18,27) or an optimum temperature (1,17) was the most important factor in nonionic surfactant solutions, because the HLB of nonionic surfactant changes with the hydrophilic chain length and/or the temperature. The same principle is applicable in ionic surfactant solutions as well. In order to change the HLB of ionic surfactants, the change of the types of counterions (33,34), the addition of salts (35), the introduction of a side chain, that is, dialkyl type surfactant (36–38), and the mixing of hydrophilic and oleophilic surfactants (35,36,39) and so on, are effective (35).

1.3.1 The Change of the Dissolution State of Ionic Surfactant + Cosurfactant with the Compositions

Continuous Change of HLB in Surfactants with Variables

Efficient surfactant molecule including cosurfactant has to be composed of strong hydrophobic (lipophilic) and lipophobic (hydrophilic) groups because the insolubility of surfactant towards water and oil is important. Because of this fact the saturation concentrations of singly dispersed species in water and hydrocarbon are both very small and most of the surfactant in solution is in an aggregated state regardless of the aggregation number being finite (micelle) or infinite (surfactant phase or liquid crystal). Although the solubility of surfactant in solvents should be small, the solubility of solvent (oil and water) in surfactant has to be large. If the solubility of water (or hydrocarbon) in the surfactant phase is infinite, that is, water (or hydrocarbon) and surfactant are mutually soluble, the surfactant disperses in water (or hydrocarbon) forming micelles. Critical solution of surfactant and solvent, that is, infinite solubility of solvent in surfactant is a necessary condition for micellar dispersion, that is, soluble surfactants (40,41). When the hydrophile–lipophile property of surfactant is just balanced, however, the solubility of water and oil in surfactant are both finite and a three-phase solution consisting of water, surfactant, and oil phases is obtained.

These situations are clearly shown in the phase diagram of a 3 wt% NaCl aq./$R_{12}SO_4Na$/glycerol mono(2 ethylhexyl)-ether/$C_{10}H_{22}$ system as a function of the composition in Fig. 1.17 (42). Water dissolves in ionic surfactant + cosurfactant phase infinitely at lower composition of cosurfactant and aqueous micellar solution, W_m, is obtained on one hand and hydrocarbon will dissolve infinitely at higher composition of cosurfactant and reversed micellar solution, O_m, is obtained on the other hand. These solubilities are expected because the surfactant is hydrophilic at a lower composition of cosurfactant and lipophilic at a higher composition of cosurfactant. Below the surfactant/oil solubility curve in Fig. 1.17, surfactant and oil phases separate and above the water/surfactant solubility curve, water and surfactant phases separate. The three-phase region, III, consisting of water, surfactant, and oil phases is observed at the intermediate composition at which the hydrophile–lipophile property of surfactant just balances. Realms W_m and O_m are one phase, because oil and water dissolve completely in micellar solution and reversed micellar solution, respectively. The solubilization is maximum just before the separation of water (or oil) phase from surfactant phase occurs. The solubility of oil (or water) in the system (actually in the surfactant phase), however, further increases and finally the solubility of oil (or water) in the surfactant phase becomes infinite. Micellar dispersion, surfactant phase separation, and reversed micellar dispersion occur continuously due to the competitive dissolution of water and oil in the surfactant phase with the change of HLB. Hence, the solubilization in aqueous and hydrocarbon media should be studied concurrently.

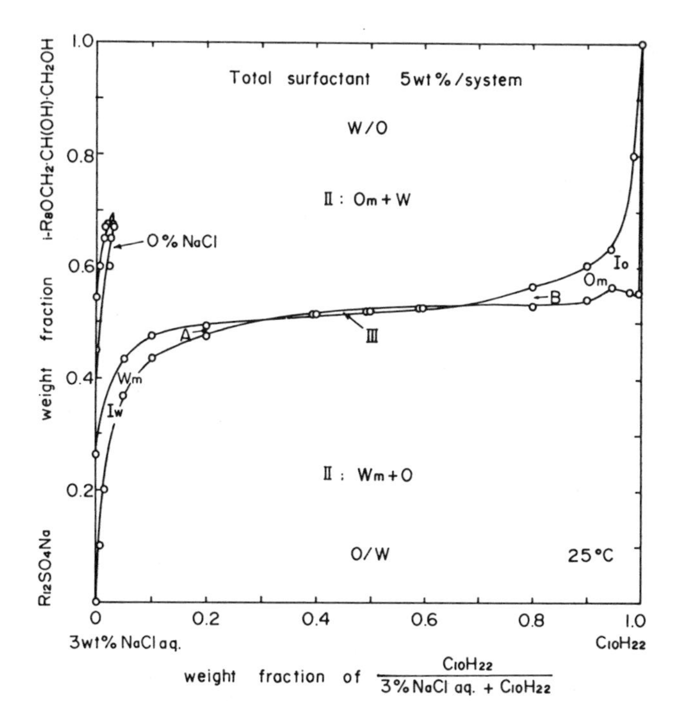

Figure 1.17 Phase diagram of a $C_{12}H_{25}SO_4Na$/glycerolmono(2-ethylhexyl)ether/3 wt% NaCl aq $C_{10}H_{22}$ system at 25°C. Total surfactant concentration is 5 wt%. A similar phase diagram was obtained by replacing NaCl with sodium pyroglutamate. [Reproduced by permission of *American Chemical Soc., J. Phys. Chem.*, 88, 5126 (1984).]

1.3.2 The Effect of the Types of Counterions, the Types of Surfactants, and Hydrocarbon Chain Length of Surfactants

Temperature change usually does not affect the HLB of ionic surfactant to a significant degree. The HLB of ionic surfactant changes not only with the composition of the mixture with other surfactants, but by the number of the hydrocarbon chain of the surfactant, the concentration of added salts, the types of counterions, the chain length of hydrocarbons (solvents), and so on (35).

Similar phase diagrams were observed in $R_{12}OCH_2$-$CH_2SO_4Ca_{1/2}$/*i*-$R_8OCH_2CH(OH)CH_2OH/H_2O/C_{10}H_{22}$, $R_{12}SO_4Na/R_8O(CH_2CH_2O)_2H$/3 wt% NaCl aqueous solu-

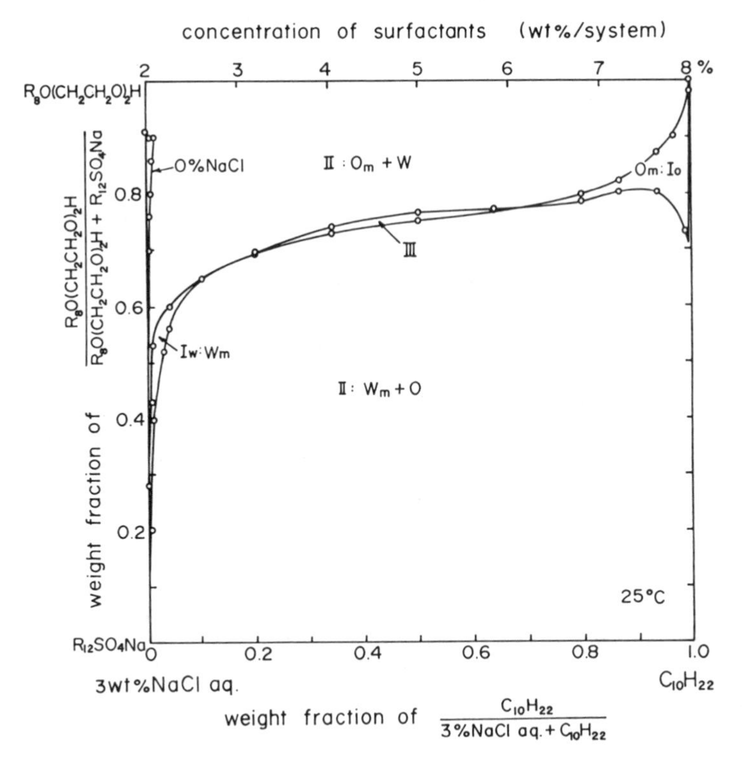

Figure 1.18 Phase diagram of a $C_{12}H_{25}SO_4Na/C_8H_{17}O(CH_2CH_2O)_2H$/3 wt% NaCl aq./$C_{10}H_{22}$ system at 25°C. Total surfactant concentration changes from 2 wt% in aqueous solution to 8 wt% in decane solution. [Reproduced by permission of *Amer. Chem. Soc., J. Phys. Chem.*, 88, 5126 (1984).]

tion/$C_{10}H_{22}$, and $R_{16}SO_4Na$/i-$R_8OCH_2CH(OH)CH_2OH$/1.0 wt% NaCl aq./liquid paraffin (Exxon's Crystol 70) at 25°C and shown in Figs. 1.18–1.20 (42).

In these systems the HLB of ionic surfactant was changed with the composition of hydrophilic surfactant and lipophilic cosurfactant. The continuous change from aqueous micellar solution to nonaqueous micellar solution via surfactant phase occurred with the composition change. The solubilization of water or oil was very large where the hydrophile–

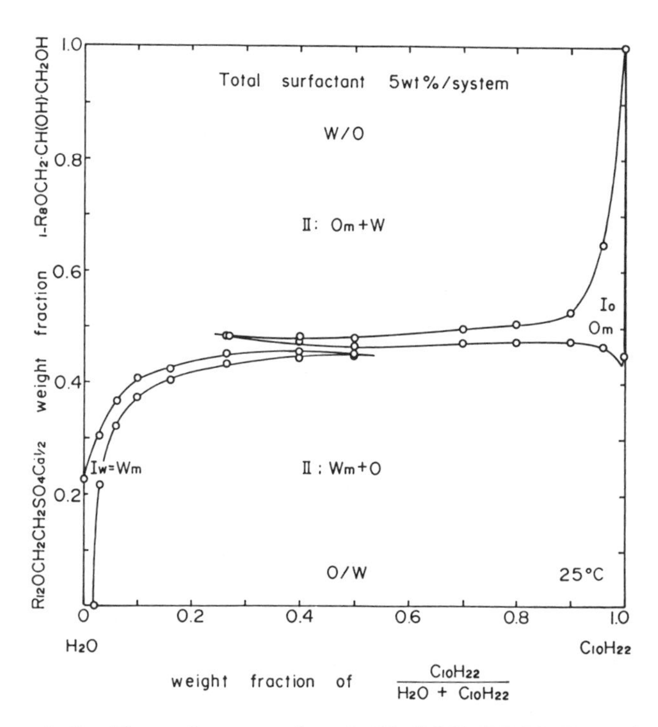

Figure 1.19 Phase diagram of a $C_{12}H_{25}OCH_2CH_2SO_4Ca_{1/2}$/glycerol mono(2-ethylhexyl)ether/H_2O/$C_{10}H_{22}$ system at 25°C. Total surfactant concentration is 5 wt%. The abscissa and ordinate are the weight fractions of solvents and surfactants, respectively, in Figs. 1.17–1.20. [Reproduced by permission of *ACS*, *J. Phys. Chem.*, 88, 5126 (1984).]

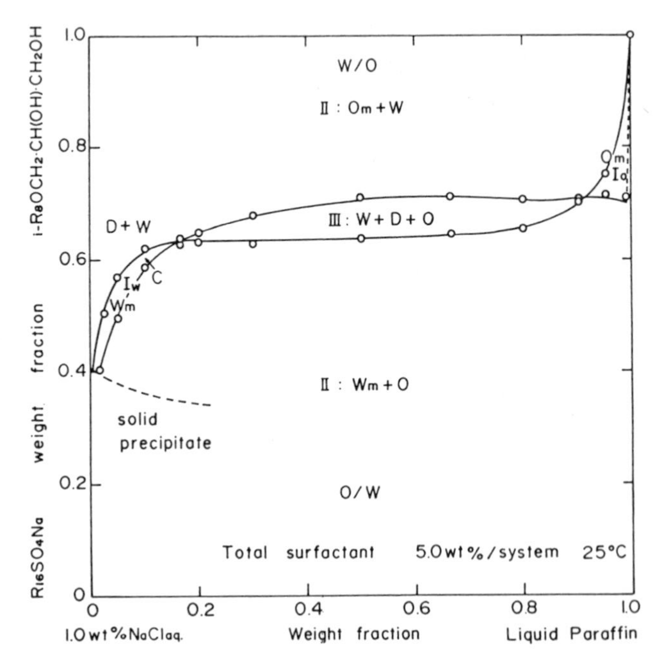

Figure 1.20 Phase diagram of a sodium hexadecylsulfate/glycerol mono(2-ethylhexyl)ether/1.0 wt% NaCl aq./liquid paraffin system at 25°C. Total surfactant concentration is 5 wt%. [Reproduced by permission of *Amer. Chem. Soc., J. Phys. Chem.*, 88, 5126 (1984).]

lipophile property of surfactant mixture balances. Hence, it is evident that both ionic and nonionic surfactant solutions could be treated in a similar manner, provided the HLBs of the two surfactants are reasonably close. It is essential to realize that structures with widely dissimilar HLBs did not give this behavior and for such couples the behavior of ionic and nonionic surfactants may appear quite different.

A system containing a calcium salt of the ionic surfactant was studied and shown in Fig. 1.19. In this system the solubilization is much larger and we could obtain a one-phase region over any water versus decane composition with 5 wt% of total surfactant per system. But, the one-phase regions I_W and I_O are separated by the appearance of a lamellar liquid crystalline phase. In the case of the Aerosol OT solution, aqueous micellar solution, I_W, continuously changed to reversed micellar solution, I_O, via the surfactant phase (43).

If the HLB of ionic surfactant is strongly hydrophilic, the addition of salt is necessary to suppress the hydrophilic property of the ionic surfactant in order to change the HLB of the surfactant mixture continuously. If salt is not added, the solubilization of oil in aqueous micellar solution is small as shown on the left-hand side of Figs. 1.17 and 1.18. Hence, if a too lipophilic cosurfactant, or a too hydrophilic ionic surfactant without salt were used, the solution behavior of ionic surfactant + cosurfactant would be different from nonionic ones and the two cases could not be treated uniformly.

Sodium hexadecyl sulfate alone does not dissolve forming micelles at 25°C, because the Krafft point is too high. Due to the mixing with cosurfactant, glycerol mono(2-ethylhexyl)-ether, the Krafft point is depressed and the surfactant dissolves at 25°C as shown in Fig. 1.20. The solubilizing power of the system is very large, because 16 wt% of liquid paraffin was solubilized in 5 wt% per system of surfactant solution. In the case of hexadecane, not less than 27 wt% was solubilized in the same surfactant solution. The value is nearly equal to that of decane in the $R_{12}SO_4Na/C_8(CH_2CH_2O)_2OH$ combination (Fig. 1.18). Since the solubilization decreases with the hydrocarbon chain length of solubilizate, we can conclude that the use of a longer chain surfactant is very effective to enhance solubilizing power.

1.3.3 The Effect of Temperature on the HLB of Ionic Surfactant and Cosurfactant Mixture

Phase equilibria are extremely sensitive to temperature change in the nonionic surfactant solutions, which means a temperature sensitive change of HLB in nonionics.

The effect of temperature on the composition of the one-phase regions in Figs. 1.17 and 1.20 is plotted in Figs. 1.21 and 1.22. In the system composed of $R_{12}SO_4Na/R_8O$-$(CH_2CH_2O)_2H$/3 wt% NaCl aq./$C_{10}H_{22}$ (Fig. 1.18), the composition of one-phase regions gradually changed with temperature. It is evident from Figs. 1.21 and 1.22, that the composition of one-phase regions, that is, aqueous micellar region, W_m, and nonaqueous micellar region, O_m, in Figs.

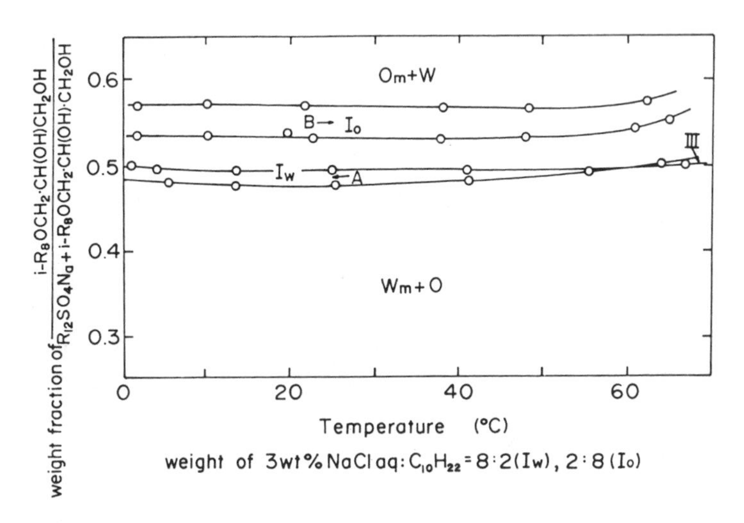

Figure 1.21 The effect of temperature on the composition of cosurfactant in one-phase regions, I_W and I_O, in the system $C_{12}H_{25}SO_4Na$/glycerol mono(2-ethylhexyl)ether/3 wt% NaCl aq./$C_{10}H_{22}$. Total surfactant concentration is 5 wt%. [Reproduced by permission of *Amer. Chem. Soc.*, *J. Phys. Chem.*, 88, 5126 (1984).]

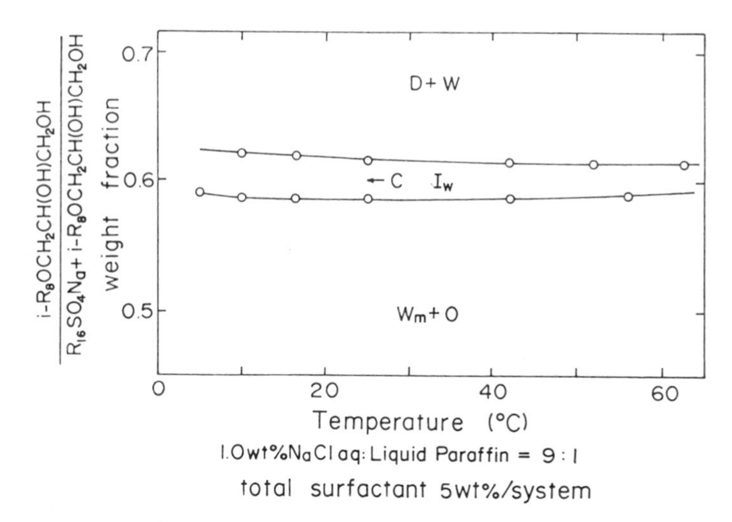

Figure 1.22 The effect of temperature on the composition of cosurfactant in a one-phase region, I_W, in the system $C_{16}H_{33}SO_4Na$/glycerol mono(2-ethylhexyl)ether/1 wt% NaCl aq./liquid paraffin. Total surfactant concentration is 5 wt%. [Reproduced by permission of *Amer. Chem. Soc.*, *J. Phys. Chem.*, 88, 5126 (1984).]

1.17 and 1.20 shows only insignificant dependence on temperature. This is a remarkable fact for many practical purposes, because the solubilizing power is strongly enhanced in these systems, and one is temperature sensitive (nonionics) and the other is temperature independent (ionics).

Lower critical solution phenomena of a water/nonionic surfactant (44,45), upper critical solution phenomena of a hydrocarbon/nonionic surfactant system (46), and the existence of a three-phase region (25) were known.

It was shown clearly that the three-phase region (water/surfactant/oil phases) appears between the two critical solution temperatures of the water/surfactant (containing oil) and oil/surfactant (containing water) phases (47–53).

In an ionic surfactant solution a similar three-phase region consisting of water, surfactant, and oil phases is observed as a function of the salt concentration of the hydrocarbon chain length of the oils (54,55) or composition of the cosurfactant (42).

REFERENCES

1. K. Shinoda and H. Saito, *J. Colloid Interface Sci.*, **26**, 70 (1968).
2. K. Shinoda, *J. Colloid Interface Sci.*, **24**, 4 (1967).
3. H. Kunieda and K. Shinoda, *Bull. Chem. Soc. Jpn.*, **55**, 1777 (1982).
4. L. M. Prince, *J. Soc. Cosmet. Chem.*, **21**, 1932 (1970).
5. C. A. Miller and L. E. Scriven, *J. Colloid Interface Sci.*, 33, 360 (1970).
6. C. A. Miller and P. Neogi, "Thermodynamics of Microemulsions: Combined Effects of Dispersion Entropy of Drops and Bending Energy of Surfactant Films," *AIChE J.*, **26**, 212 (1980).
7. E. Ruckenstein, *J. Dispersion Sci. Technol.*, **2**, 1 (1980).
8. A. M. Bellocq, D. Bourbon, and B. Lemanceau, *J. Dispersion Sci. Technol.*, **2**, 27 (1981).
9. J. Biais, P. Bothorel, B. Clin, and P. Lalanne, *J. Dispersion Sci. Technol.*, **2**, 67 (1981).
10. D. J. Mitchell and B. W. Ninham, *J. Chem. Soc. Faraday Trans.* II, **77**, 601–29 (1981).
11. K. Shinoda, *Progr. Colloid Polymer Sci.*, **68**, 1 (1983).
12. C. T. Murphy, Thesis, University of Minnesota (1966).
13. S. Friberg and I. Lapczynska, *Progr. Colloid Polymer Sci.*, **56**, 16 (1975).

14. S. Friberg, I. Lapczynska, and G. Gillberg, *J. Colloid Interface Sci.*, **56**, 19 (1976).

15. G. Gillberg, L. Eriksson, and S. Friberg, in *Emulsions, Lattices and Dispersion* (Becher and Yudenfreund, Eds.), Marcel Dekker, New York (1978), p. 201.

16. H. Saito and K. Shinoda, *J. Colloid Interface Sci.*, **24**, 10 (1967).

17. L. Rayleigh, *Phil. Mag.*, **41**, 107, 274, 447 (1871).

18. K. Shinoda and T. Ogawa, *J. Colloid Interface Sci.*, **24**, 56 (1967).

19. K. Shinoda and H. Arai, *J. Colloid Interface Sci.*, **25**, 429 (1967).

20. J. T. Davis, in *Recent Progress in Surface Science*, Vol. 2 (J. F. Danielli, K. G. A. Parkhurst, and A. C. Riddiford, Eds.), Academic, New York (1964), p. 129.

21. P. Ekwall, L. Mandell, and K. Fontell, *Mol. Cryst. Liq. Cryst.*, **8**, 157 (1969).

22. G. Gillberg, H. Lehtinen, and S. Friberg, *J. Colloid Interface Sci.*, **33**, 40 (1970).

23. S. Friberg, P. O. Jansson, and E. Cederberg, *J. Colloid Interface Sci.*, **55**, 614 (1976).

24. S. Friberg, L. Mandell, and K. Fontell, *Acta Chem. Scand.*, **23**, 1055 (1969).

25. K. Shinoda and H. Arai, *J. Phys. Chem.*, **68**, 3485 (1964).

26. H. Saito and K. Shinoda, *J. Colloid Interface Sci.*, **32**, 647 (1970).

27. K. Shinoda and H. Kunieda, *J. Colloid Interface Sci.*, **42**, 384 (1973).

28. P. A. Winsor, *Trans. Faraday Soc.*, **44**, 376 (1948).

29. P. A. Winsor, *Solvent Properties of Amphilic Compounds*, Butterworth, London (1954).

30. J. H. Schulman and D. P. Riley, *J. Colloid Sci.*, **3**, 383 (1948).

31. J. H. Schulman and J. A. Friend, *J. Colloid Sci.*, **4**, 497 (1949).
32. J. H. Schulman and J. B. Montagne, *Ann. N. Y. Acad. Sci.*, **92**, 366 (1961).
33. K. Shinoda and T. Hirai, *J. Phys. Chem.*, **81**, 1842 (1977).
34. K. Shinoda, M. Hanrin, H. Kunieda, and H. Saijo, *Colloid Surfaces*, **2**, 301 (1981).
35. K. Shinoda, *Pure Appl. Chem.*, **52**, 1195 (1980).
36. H. Sagitani, T. Suzuki, M. Nagai, and K. Shinoda, *J. Colloid Interface Sci.*, **87**, 11 (1982).
37. A. Graciaa, Y. Barakat, M. El-Emary, L. Fortney, R. Schechter, S. Yiv, and W. H. Wade, *J. Colloid Interface Sci.*, **89**, 209 (1982).
38. K. Shinoda and H. Sagitani, *J. Phys. Chem.*, **87**, 2018 (1983).
39. K. Shinoda, H. Kunieda, N. Obi, and S. E. Friberg, *J. Colloid Interface Sci.*, **80**, 304 (1981).
40. K. Shinoda, *J. Colloid Interface Sci.*, **34**, 278 (1970).
41. K. Shinoda, *J. Phys. Chem.*, **85**, 3311 (1981).
42. K. Shinoda, H. Kunieda, T. Arai, and H. Saijo, *J. Phys. Chem.*, **88**, 5126 (1984).
43. H. Kunieda and K. Shinoda, *J. Colloid Interface Sci.*, **75**, 601 (1980).
44. R. R. Balmbra, J. S. Clunie, J. M. Corkill, and J. F. Goodman, *Trans. Faraday Soc.*, **52**, 1661 (1962).
45. K. Shinoda, *J. Colloid Interface Sci.*, **34**, 278 (1971).
46. K. Shinoda and H. Arai, *J. Colloid Sci.*, **20**, 93 (1965).
47. M. L. Robbins, in Micellization, Solubilization and Microemulsions (K. L. Mittal, Ed.), Plenum Press, New York (1977).
48. M. Bourrel, Ch. Koukounis, R. Schechter, and W. Wade, *J. Dispersion Sci. Technol.*, **1**, 13 (1980).

49. H. Kunieda and S. E. Friberg, *Bull. Chem. Soc., Jpn.*, **54**, 1010 (1981).
50. C. U. Herrman, G. Klar, and M. Kahlweit, *J. Colloid Interface Sci.*, **82**, 6 (1981).
51. M. Kahlweit, *J. Colloid Interface Sci.*, **90**, 197 (1982).
52. H. Kunieda and K. Shinoda, *J. Dispersion Sci. Technol.*, 3, 233 (1982).
53. M. Kahlweit, E. Lessner, and R. Strey, *J. Phys. Chem.*, 87, 5032 (1983).
54. R. N. Healy and R. L. Reed, *Soc. Pet. Eng. J.*, **14**, 491 (1974).
55. W. H. Wade, J. C. Morgan, R. S. Schechter, J. K. Jacobson, and J. L. Salazer, *Soc. Pet. Eng. J.*, 18, 242 (1978).

CHAPTER 2

Concepts of HLB, HLB Temperature, and HLB Number

2.1 CONCEPT OF HYDROPHILE–LIPOPHILE BALANCE, HLB, OF SURFACTANT

This concept is most important in selecting a suitable surfactant that will satisfactorily emulsify or solubilize given ingredients at a given temperature. This means that the hydrophile–lipophile balance (HLB) of the surfactant is the most useful concept in the studies of emulsions, solubilization, microemulsions, and many other applications. Clayton (1) has drawn attention to the concept of balanced surfactants embodied in a series of patents dating back to 1933. In a given homologous series of surfactants, such as polyoxyethylene alkyl(aryl)ethers, there is a point in the range at which hydrophile–lipophile balance is optimum for the particular application.

Schemes designed to put this concept on a quantitative basis have been advanced: the HLB numbers of Griffin (2,3) and Davies (4), the H/L numbers of Moore and Bell (5), the water number of Greenwald et al. (6), the HLB temperature (or PIT, phase inversion temperature) of Shinoda and coworkers (7–11) and EIP (emulsion inversion point) of Marszall (12–14) for nonionic surfactant, and the HLB composition of Shinoda et al. (15–17) for ionic surfactant may be mentioned.

Among the first three mentioned formulas, the HLB number by Griffin, which has been designated the HLB (number) method, is the most widely used. However, an HLB number is a number assigned to a molecule as such without any consideration of the properties of the two solvents present. This is unfortunate since the actual HLB of adsorbed surfactant at the oil/water interface changes with the types of oil, temperature, additives in oil and water phases, and so on.

In contrast to these approaches, the HLB temperature (or PIT) method by Shinoda (a) adopts a characteristic property for the emulsion as an indicator of the HLB of surfactant used and (b) utilizes the information obtained from the effect of variables on the properties of the system in a systematic manner. Since PIT (HLB temperature) is a characteristic property of the emulsion with the surfactant present, the effect of additives on the solvents, the effect of the mixing of emulsifiers or of mixing of oils, and so on, and the effect of the structure of molecules are all reflected on the PIT and tell us how the in situ HLB of the emulsifier at the interface is changed. The important concept of: (a) the effect of the temperature on the HLB and (b) the effect of solvent on the HLB, and so on, were established (7,10) from the PIT (HLB temperature) studies. The basis of the system is the fact that the true HLB of an emulsifier is a function not only of the structure and composition of surfactant, but also of the types of oils, temperature, and so on.

The HLB number and HLB temperature are interrelated and convertible with each other, but specifications of the system, that is, the kinds of oils, temperature, and so on, are necessary. The studies of phase diagrams, dispersion types, emulsion stability, and so on, in a water/oil/emulsifier system, varying the hydrophilic–lipophilic balance of the emulsifier, the temperature, or employing different mixing ratios of emulsifiers, including ionic surfactants, give a clear understanding of emulsions.

In order to evaluate an HLB of an ionic surfactant or to evaluate the optimum mixing ratio of an ionic surfactant and cosurfactant, a new method is proposed and explained in a later section of this chapter. Table 2.1 illustrates the relationships among various approaches.

Since any method will treat hydrophile–lipophile balance,

TABLE 2.1 The Relations in the HLB(T) System

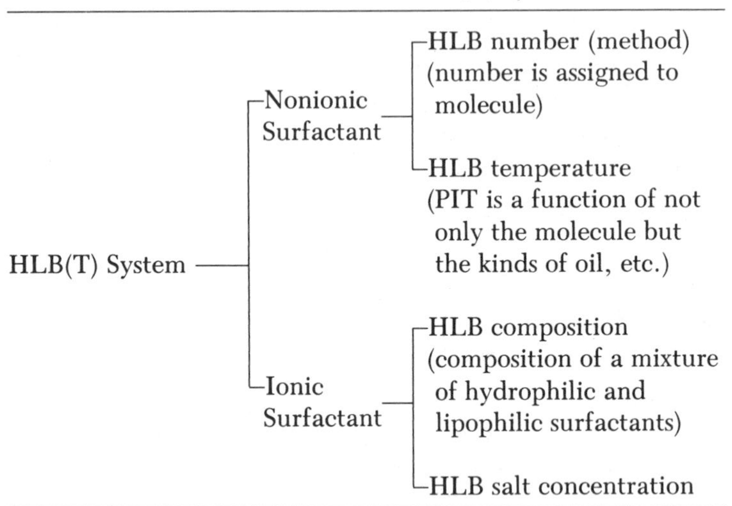

HLB, it is suggested to designate the whole system as an HLB(T) system since the important effect of temperature is taken into account.

2.2 HLB TEMPERATURE (HYDROPHILE–LIPOPHILE BALANCE TEMPERATURE OR PIT)

It is clear from Chapter 1 that the HLB of a nonionic surfactant is a function of temperature, because the interaction between water and a hydrophilic group or oil and a lipophilic group of surfactant changes with temperature (18–22). *The required HLB numbers of oils and the HLB numbers of surfactants* are determined from the maximum stability of emulsions as a function of the HLB of surfactants at 25°C.

However, it is not easy to determine exact HLB numbers, because the emulsion stability does not change significantly with small variations of the HLB numbers of surfactants as shown in Fig. 2.1 (9).

The abscissa is the PITs, that is, HLB numbers of a series of polyoxyethylene nonylphenylethers. Emulsifiers, the PITs of which are in the range 70 ± 20°C (HLB number 13 ± 1) higher than the storage temperature yield more stable emulsions (11).

The situation is explained by Griffin as follows:

> In its present form, the HLB (number) lacks exactness. A suitable simple laboratory method of measuring HLB numbers of surfactants accurately is needed. We have tried a variety of methods including solubility in water, or various solvents, ratio of solubility in two solvents, solubilization behavior both for oils and dyes, surface and interfacial ten-

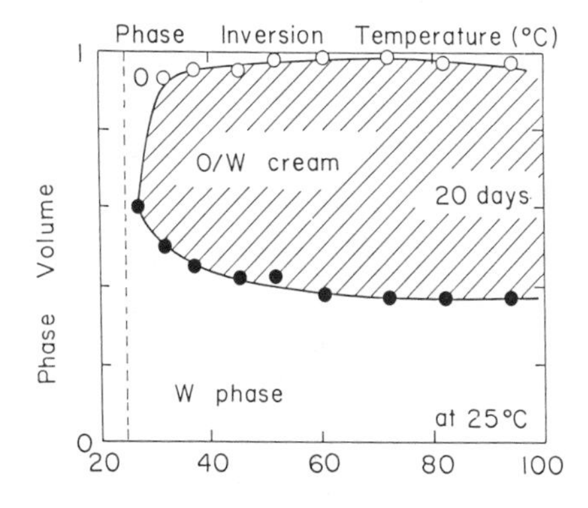

Figure 2.1 The effect of the phase inversion temperature (PIT) of emulsifiers on the volume fraction of oil, cream, and water phases 20 days after emulsification. (Emulsified and stored at 25°C using a series of polyoxyethylene nonylphenylethers, the PITs of which vary from 27–94°C.) [Reproduced by permission of Academic, *J. Colloid Interface Sci.*, *30*, **258** (1969).]

> sion data, cloud point behavior, and many other properties. Of these, the most promising is the determination of the cloud point of an aqueous solution of the surfactant. However, this test still possesses several limitations (2).

The cloud point of an aqueous solution of a surfactant accurately reflects the hydrophilic property of the surfactant. The cloud points in the presence of oils more closely reflect the HLB of surfactant for given oils. If we adopt a characteristic property of emulsions, such as phase inversion temperature (PIT), as a measure of the HLB, it is much more accurately determined (±1–2°C) and naturally reflects any experimental condition, such as the effect of the sizes and types of hydrophilic groups and lipophilic groups of the surfactants. In addition, information is provided on the effect of the concentration of emulsifier, the phase volume, the additives in both water and oil phases, the temperature, the effect of blending of oils and surfactants, the molecular structure, and so on (7–11, 8–22).

The comparison between the HLB temperature and the HLB number with regard to the information of various factors is summarized in Table 2.2 (10). Hence, it is evident that the PIT is one of the most suitable parameters for measuring the HLB of nonionic surfactants (7,8,18). The HLB number assigned to a surfactant molecule, in principle, cannot give any information on any additional variable. Empirical corrections such as the "required HLB numbers of oils," "required HLB numbers of water containing salts," and so on, are necessary based on the stability measurement and by the aid of the PIT data. On the other hand, the PIT is not observed if a surfactant is too lipophilic. In such a case, PIT has to be evaluated by extrapolation from more hydrophilic homologs, or the optimum oxyethylene chain length for W/O-type emulsions (whose PITs are lower than 0°C) is

TABLE 2.2 The Comparison between the PIT (HLB Temperature) and the HLB Number in Regard to the Information on Various Factors[a]

PIT (HLB Temperature)	Factors	HLB Number
○	Hydrophile–lipophile balance of surfactant	●
○	Types of hydrophilic moiety of surfactant	▲
○	Types of lipophilic moiety of surfactant	▲
○	Types of oils	●
○	Additives in water and/or oil phase	▲
○	Concentration of emulsifier	☆
○	Phase volume	☆
○	Temperature	☆
○	Emulsion types	●
○	Correlation with the other properties	▲
◑	In the case of ionic surfactant	▲

[a]Key: ○, Accurate information is available; ●, Less accurate information is available; ▲, Crude information is available; ☆, Almost no information is available; ◑, Information is available in the presence of salt and cosurfactant.

evaluated from the empirical stability/PIT/oxyethylene chain length relation (11).

Although the HLB temperature gives all the information about the experimental conditions in *nonionics*, the addition of an optimum amount of salt and cosurfactant is necessary to observe the phase inversion in emulsion in *ionic* surfactant combinations, since the HLB of ionic surfactant alone does not change appreciably with temperature and PIT is usually not observed. The effect of temperature on the emulsion type in ionic surfactant solution seems opposite to that in nonionics. It is the O/W-type emulsion at a higher temperature and the W/O-type emulsion at a lower temperature in a solution of Aerosol OT (23). The optimum mixing ratios of ionic surfactants with cosurfactants to form a lamellar liquid crystal also gives relatively accurate information concerning the HLB of ionic surfactants (17).

Emulsions stabilized with nonionic agents are W/O types at high temperature but change to an O/W type at low temperature. This phenomenon is evidence that the hydrophile–lipophile balance of nonionic surfactant changes appreciably to more hydrophilic with temperature depression. The hydration forces between the hydrophilic moiety of surfactant and water are stronger at the lower temperature so that the surfactant is more hydrophilic, and the adsorbed monolayer at an oil/water interface may have a convex curvature towards water. When the hydration between the hydrophilic moiety of the surfactant and water diminishes at the higher temperature, the surfactant becomes more lipophilic. Hence, the adsorbed monolayer at an oil/water interface may have concave curvature towards water (8) and the inversion of an emulsion type occurs.

The PIT of an emulsion is the temperature at which the hydrophilic and lipophilic property of a nonionic surfactant balances, and the hydrophile–lipophile balance of a surfactant changes with temperature. Hence, a correlation between the HLB temperature and the HLB number is expected. This relation is shown in Fig. 2.2. We can determine the HLB number from the HLB temperature of a surfactant by the aid of the calibration curve in Fig. 2.2. As the PIT

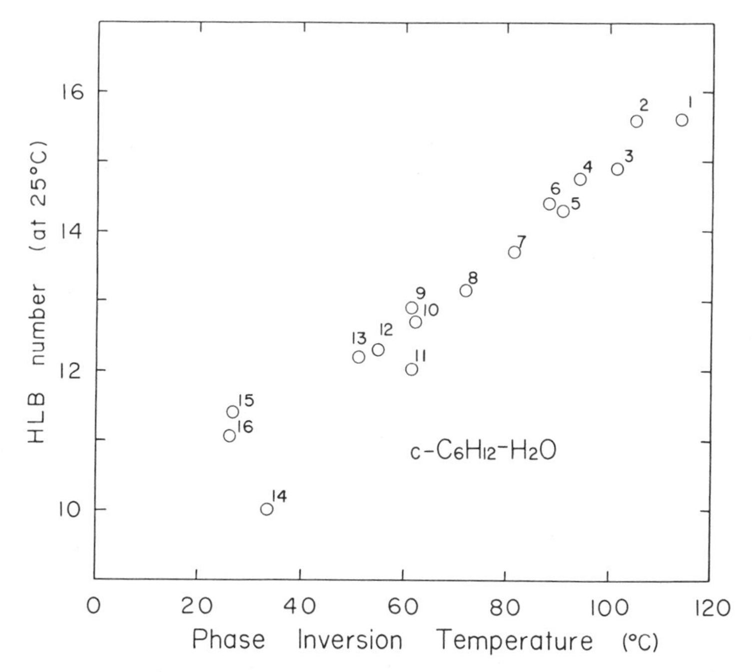

Figure 2.2 The correlation between the HLB numbers of nonionic surfactants and the PIT in cyclohexane/water emulsions stabilized with the surfactants (3 wt%). Key: 1, Tween 40; 2, $i\text{-}R_9C_6H_4O(CH_2CH_2O)_{17.7}H$; 3, Tween 60; 4, $i\text{-}R_9C_6H_4O(CH_2CH_2O)_{14.0}H$; 5, $i\text{-}R_{12}C_6H_4O(CH_2CH_2O)_{15}H$; 6, $R_{12}O(CH_2CH_2O)_{10.8}H$; 7, $i\text{-}R_8C_6H_4O(CH_2CH_2O)_{10}H$; 8, $i\text{-}R_9C_6H_4O(CH_2CH_2O)_{9.7}H$; 9, $i\text{-}R_8C_6H_4O(CH_2CH_2O)_{8.6}H$; 10, $i\text{-}R_{12}C_6H_4O(CH_2CH_2O)_{9.7}H$; 11, $R_{12}O(CH_2CH_2O)_{6.3}H$; 12, $i\text{-}R_{12}C_6H_4O(CH_2CH_2O)_{9.4}H$; 13, $i\text{-}R_9C_6H_4O(CH_2CH_2O)_{7.4}H$; 14, $R_{12}O(CH_2CH_2O)_{4.2}H$; 15, $i\text{-}R_8C_6H_4O(CH_2CH_2O)_6H$; 16, $i\text{-}R_9C_6H_4O(CH_2CH_2O)_{6.2}H$.

changes markedly with the types of oils, the correlation between the HLB numbers and the PITs in different oil/water emulsions differ from each other as shown in Fig. 2.3. Since the HLB number of a surfactant at different oil/water interfaces differs, a correction for the types of oils is necessary in the HLB-number system, which is known as "required HLB

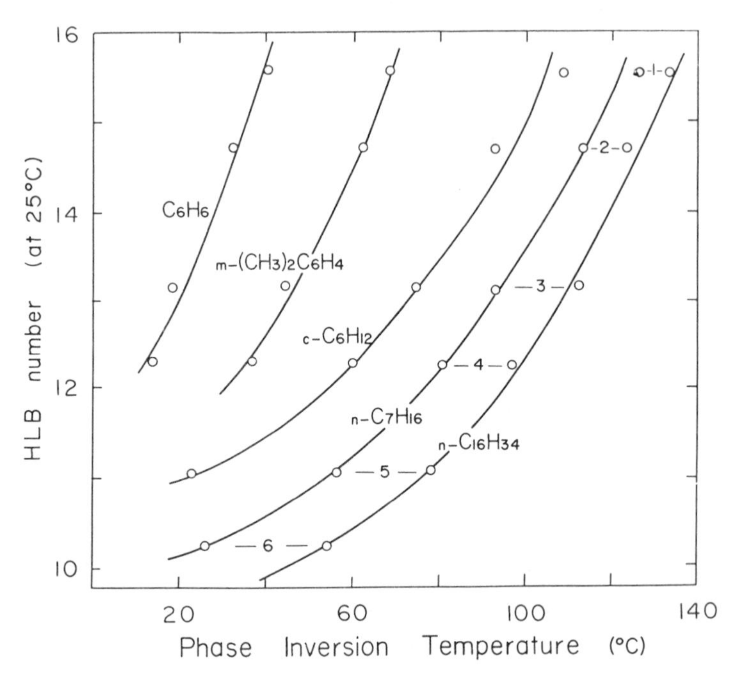

Figure 2.3 The correlation between the HLB numbers and the PITs in various oil/water (1:1) emulsions stabilized with nonionic surfactants (1.5 wt%). Key: 1, $i\text{-}R_9C_6H_4O(CH_2CH_2O)_{17.7}H$; 2, $i\text{-}R_9C_6H_4O\text{-}(CH_2CH_2O)_{14.0}H$; 3, $i\text{-}R_9C_6H_4O(CH_2CH_2O)_{9.6}H$; 4, $i\text{-}R_9C_6H_4O\text{-}(CH_2CH_2O)_{7.4}H$; 5, $i\text{-}R_9C_6H_4O(CH_2CH_2O)_{6.2}H$; 6, $i\text{-}R_9C_6H_4O\text{-}(CH_2CH_2O)_{5.3}H$. [Reproduced by permission of Chem. Soc. Japan, *J. Chem. Soc.* Japan, 89, 435 (1978).]

numbers of oils (24)." The HLB temperature, on the other hand, reflects the types of oils and no correction is needed.

The HLB number is a number assigned at 25°C but is a function of temperature. Since the effect of temperature on the HLB numbers of ionic and nonionic surfactants is different, the same HLB numbers at 25°C for ionics and nonionics mean similar HLB numbers for ionics and lower HLB numbers for nonionics at a higher temperature, and vice versa.

The HLB number of a nonionic surfactant whose PIT (HLB temperature) = 25°C is 9.5 for liquid paraffin and 11.1 for cyclohexane (11). Figure 2.3 is useful to find the "required HLB numbers of oils at 25°C," because the HLB numbers of surfactants whose HLB temperatures are 25–70°C higher than storing temperatures are the "required HLB numbers" (9). The required HLB numbers estimated from the HLB-temperature system and those recommended by the Atlas pamphlet (24) are compared in Table 2.3.

It is evident that both systems are complementary to each other. Information obtained from both systems can be utilized. Griffin proposed his way to determine the HLB number of surfactant. We propose a modified way for that based on the solution properties and inversion in emulsion types in surfactant/water/oil systems.

TABLE 2.3 The Comparison between the Atlas Required HLB Numbers and the HLB Numbers at 25°C Estimated from the PIT

Types of Oils	HLB Numbers (Atlas) (24)	HLB Numbers (PIT) (10)	PIT (°C)
Mineral oil (paraffinic)	10	10	110
Propene tetramer	14	12	—
Kerosene	14	12	94
Trichlorotrifluoroethane	14	12.5	—
Cyclohexane	15	13	70
Carbontetrachloride	16	13.5	53
Xylene	14	14.5	46
Toluene	15	15.5	38
Benzene	15	16.5	21

Correlation Among the PIT (HLB temperature) and the Number of Methylene Groups and Oxyethylene Groups in Surfactant

Due to Griffin the HLB number is given by the weight fraction of oxyethylene plus polyhydric alcohol groups in a surfactant molecule. Suppose the HLB number of emulsifier in an emulsion of a certain oil/water system is 10.0 (9.5 for paraffinic liquid) at the PIT = 25°C we obtain:

$$\text{HLB number} = \frac{W_E}{W_L + W_{EO}} \times 20 = 10 \tag{1}$$

where W_L and W_{EO} are the weights of lipophilic and oxyethylene groups in the molecule, respectively.

Homologous series of emulsifiers of different hydrocarbon and oxyethylene chain lengths whose PITs are 25°C should have the same HLB number so that we obtain:

$$\text{HLB number} = \frac{W_{EO} + W'_{EO}}{W_L + W'_L + W_{EO} + W'_{EO}} \times 20 = 10.0 \tag{2}$$

where W'_L and W'_{EO} are the weight increases of hydrocarbon and oxyethylene chains in surfactant molecules, respectively. From Eqs. (1) and (2):

$$\frac{W'_{EO}}{W'_L + W'_{EO}} = \frac{1}{2} \tag{3}$$

This relation tells us that the PITs may be equal if the increases in weight of the hydrocarbon portion, W'_L, and oxyethylene group are equal when the HLB number of the emulsifier is 10 at PIT = 25°C. In other words, the PIT will remain unchanged at 25°C if 3.15 methylene groups are added per one addition of oxyethylene group.

The molecular formula of emulsifiers and PITs in emulsions composed of H_2O and C_6H_{12} are summarized in Table 2.4 (25). We can see from Table 2.4 that the hydrophile–lipophile property of surfactants at the water/cyclohexane interface balance, if 2–2.7 methylene groups are added per one oxyethylene group at 25°C.

Since the HLB number of an emulsifier whose PIT = 25°C is ≈11.6 (11),

$$\frac{W'_{EO}}{W'_L + W'_{EO}} \times 20 = 11.6$$

or

$$\frac{W'_L}{W'_{EO}} = \frac{20}{11.6} - 1 = \frac{8.4}{11.6}$$

TABLE 2.4 The Molecular Formula of Emulsifiers and PITs in Emulsions (Water/Cyclohexane) (25)

Emulsifier	PIT (°C)	HLB Numbers[a]
$R_6C_6H_4O(CH_2CH_2O)_{7.5}H$	52	13.0
$R_9C_6H_4O(CH_2CH_2O)_{8.6}H$	50	12.4
$R_{12}C_6H_4O(CH_2CH_2O)_{9.7}H$	51	12.2
$R_{16}C_6H_4O(CH_2CH_2O)_{12.4}H$	48	12.6
$R_8O(CH_2CH_2O)_{4.3}H$	25	11.9
$R_{12}O(CH_2CH_2O)_{5.8}H$	25	11.6
$R_9C_6H_4O(CH_2CH_2O)_{6.2}H$		11.1
$R_9C_6H_4O(CH_2CH_2O)_{4.5}H$	25	9.5 (liquid paraffin)

[a]These HLB numbers are assigned to an emulsifier at room temperature. Actual HLB number at 50°C is about one unit smaller.

or

$$\frac{W'_{\mathrm{L}}\,(8.4)}{W'_{\mathrm{EO}}\,(11.6)} = \frac{N'_{\mathrm{CH_2}} \times 2.28}{N'_{\mathrm{EO}}} = \frac{8.4}{11.6} \tag{4}$$

where N is the number of lower index groups. The PIT will remain unchanged at 25°C if 2.3 methylene groups are added per one addition of oxyethylene group. This conclusion is in good agreement with the relation observed in Table 2.4 (25).

2.3 HLB NUMBER

Becher, in his monumental book *Emulsions: Theory and Practice* (26) thoroughly described the HLB number, H/L number, and water number that express the hydrophile–lipophile balance of nonionic emulsifiers.

This approach to the problem due to Griffin (2) has been designated the HLB method. (HLB number method may be a more appropriate name.) In this method, an HLB number is assigned to each surfactant and is related by a scale to suitable applications. Table 2.5 shows the HLB number range

TABLE 2.5 HLB-Number Ranges and Their Application (24)

HLB-Number Range (Griffin)	(Moore and Bell)	Application
3–6	7.7	W/O Emulsifier
7–9	13.4	Wetting agent
8–18	11.1–15.9	O/W Emulsifier
13–15		Detergent
15–18	16.5	Solubilizer

required for various systems (24). As can be seen, only those materials with HLB numbers in the range of 3–6 are suitable as emulsifiers for W/O-type emulsions, while only those with HLB numbers in the range 8–18 are suitable for the preparation of O/W-type emulsions. Agents with HLB numbers in different ranges, while possessing important surface-active properties, cannot (according to this classification) be employed as emulsifying agents.

The original method of determining HLB numbers involves a long and laborious experimental procedure (2). Griffin (3) has also developed equations that permit the calculation of the HLB numbers for certain types of nonionic agents, in particular, polyoxyethylene derivatives of fatty alcohols and polyhydric alcohol fatty acid esters, including those of polyglycols.

The formulas for determining HLB numbers may be based either on analytical or composition data. For most polyhydric alcohol fatty acid esters approximate values may be calculated with the aid of the relation

$$\mathrm{HLB} = 20\left(1 - \frac{S}{A}\right) \tag{5}$$

where S is the saponification number of the ester and A is the acid number of *the acid*.

Example A: **Atmul 67, glyceryl monostearate (soap free)**

S = saponification number, 161

A = acid number of fatty acid, 198

$$\mathrm{HLB} = 20\left(1 - \frac{161}{198}\right) = 3.8$$

EXAMPLE B: Tween 20, polyoxyethylene sorbitan monolaurate

S = saponification number, 45.5 (mid-point)

A = acid number of fatty acid, 276

$$\mathrm{HLB} = 20\left(1 - \frac{45.5}{276}\right) = 16.7$$

Unfortunately, for many fatty acid esters it is difficult to get good saponification number data of, for example, lanolin and esters of tall oil and rosin beeswax. For these, Griffin gives the relation

$$\mathrm{HLB} = \frac{E + P}{5} \tag{6}$$

where E is the weight percentage of the oxyethylene content and P is the weight percentage of the polyhydric alcohol content.

EXAMPLE: **Atlas G-1441, polyoxyethylene sorbitol lanolin derivative**

E = weight percentage of oxyethylene content, 65.1

P = weight percentage of polyhydric alcohol content, 6.7

$$\mathrm{HLB} = \frac{65.1 + 6.7}{5} = 14$$

In products where only ethylene oxide is used as the hydrophilic portion, and for ethylene oxide condensation products of fatty alcohols Eq. (6) may be reduced to

$$\mathrm{HLB} = \frac{E}{5} \tag{7}$$

where E has the same meaning as Eq. (6).

EXAMPLE: **Myrj 49, polyoxyethylene stearate**

E = weight percentage of oxyethylene content, 76

$$\mathrm{HLB} = \frac{76}{5} = 15$$

These equations cannot be used for nonionic surface-active materials containing propylene oxide, butylene oxide, nitrogen, sulfur, and so on, nor can they be used for ionic agents. In these cases, the laborious experimental method (2) must be used. Phase inversion temperature (HLB temperature) in an emulsion may be determined in the case of nonionic emulsifiers containing propylene oxide, nitrogen, sulfur, and so on, so that the HLB number is obtained provided the PIT exists for a given system (7).

2.3.1 H/L Number

Moore and Bell (5) have introduced another method of calculating the hydrophile–lipophile balance of a series of emulsifying agents of the polyoxyethylene type. The hydrophobic groups included in this method are derived from fatty alcohols, saturated and unsaturated fatty acids, alkyl phenols, and castor oil. The number used to classify these agents is calculated from

$$\mathrm{H/L} = \frac{\text{number EO units} \times 100}{\text{number C atoms in lipophile}} \tag{8}$$

The range of numbers obtained and their areas of application are summarized in Table 2.6. As can be seen, the H/L numbers are of a different order of magnitude than the HLB numbers described in the previous section. Also, the method of calculation based on Eq. (8) does not apparently assign any hydrophilic value to the carboxyl group of fatty acids. Thus, a polyglycol ester of stearic acid and a polyoxyethylene derivative of stearyl alcohol, containing an equal number of ethylene oxide units, are assumed to have the same hydrophile–lipophile balance, in disagreement with the results by Griffin (27). By assigning suitable equivalent EO units to respective hydrophilic groups, the disagreement between the two methods may be reconciled.

$$\text{H/L number} = \frac{N_{\text{EO}}}{N_{\text{CH}_2}} \times 100 = \frac{W_{\text{EO}}}{W_{\text{CH}_2}} \times \frac{14}{44} \times 100 = \frac{W_{\text{H}}}{W_{\text{L}}} \times 31.8 \quad (9)$$

TABLE 2.6 H/L Ranges and Their Application According to Moore and Bell (5)

Nature of Agent	H/L	HLB Number	Application
Strongly lipophilic	20	7.7	W/O—for paraffins
Moderately lipophilic	40	11.1	O/W—for paraffins
Balanced	65	13.4	Wetting, dispersing
Moderately lipophilic	125	15.9	O/W—for polar substances
Strongly hydrophilic	150	16.5	Solubilizing

and

$$\text{HLB number} = \frac{W_H}{W_L + W_H} \times 20 \tag{10}$$

From Eqs. (9) and (10)

$$\text{HLB number} = \frac{W_H}{W_L} \times \frac{20}{1 + W_H/W_L} \approx \frac{H}{L}\,\text{number} \times \frac{20}{31.8 + \text{H/L number}} \tag{11}$$

where N_{EO}, N_{CH_2}, W_{EO}, W_H, W_L are the number of ethylene oxide units in the molecule, the number of methylene groups in the lipophile portion, the weight of ethylene oxide, the weight of the hydrophile portion, and the weight of the lipophile portion, respectively. The HLB number calculated from the H/L number by the aid of Eq. (11) is also shown in Table 2.6. Equation (11) implies that any information obtained by the HLB number or the H/L number method is combined and used. Information from HLB temperatures also complement each other.

The assigned HLB numbers at 25°C for a large number of commercial emulsifiers have been determined or calculated by the more complex methods described in this section (3). Table 2.7 gives these values arranged in order of increasing HLB numbers.

In addition to the materials listed in Table 2.7, Griffin (27) has indicated that the HLB numbers for such well-known emulsifying agents as potassium, sodium, and triethanolamine oleates are 20.0, 18.0, and 12.0, respectively. Oleic acid has an HLB number of approximately unity.

TABLE 2.7 HLB Values for Commercial Emulsifiers (3)

Name	Manufacturer[a]	Chemical Designation	Type[b]	HLB[c]
Span 85	1	Sorbitan trioleate	N	1.8
Arlacel 85	1	Sorbitan trioleate	N	1.8
Atlas G-1706	1	Polyoxyethylene sorbitol beeswax derivative	N	2
Span 65	1	Sorbitan tristearate	N	2.1
Arlacel 65	1	Sorbitan tristearate	N	2.1
Atlas G-1050	1	Polyoxyethylene sorbitol hexastearate	N	2.6
Emcol EO-50	2	Ethylene glycol fatty acid ester	N	2.7
Emcol ES-50	2	Ethylene glycol fatty acid ester	N	2.7
Atlas G-1704	1	Polyoxyethylene sorbitol beeswax derivative	N	3
Emcol PO-50	2	Propylene glycol fatty acid ester	N	3.4
Atlas G-922	1	Propylene glycol monostearate	N	3.4
Pure	6	Propylene glycol monostearate	N	3.4
Atlas G-2158	1	Propylene glycol monostearate	N	3.4

Emcol PS-50	2	Propylene glycol fatty acid ester	N	3.4
Emcol EL-50	2	Ethylene glycol fatty acid ester	N	3.6
Emcol PP-50	2	Propylene glycol fatty acid ester	N	3.7
Arlacel C	1	Sorbitan sesquioleate	N	3.7
Arlacel 83	1	Sorbitan sesquioleate	N	3.7
Atlas G-2859	1	Polyoxyethylene sorbitol 4.5 oleate	N	3.7
Atmul 67	1	Glycerol monostearate	N	3.8
Atmul 84	1	Glycerol monostearate	N	3.8
Tegin 515	5	Glycerol monostearate	N	3.8
Aldo 33	4	Glycerol monostearate	N	3.8
Pure	6	Glycerol monostearate	N	3.8
Atlas G-1727	1	Polyoxyethylene sorbitol beeswax derivative	N	4
Emcol PM-50	2	Propylene glycol fatty acid ester	N	4.1
Span 80	1	Sorbitan monooleate	N	4.3
Arlacel 80	1	Sorbitan monooleate	N	4.3
Atlas G-917	1	Propylene glycol monolaurate	N	4.5

TABLE 2.7 (Continued)

Name	Manufacturer[a]	Chemical Designation	Type[b]	HLB[c]
Atlas G-3851	1	Propylene glycol monolaurate	N	4.5
Emcol PL-50	2	Propylene glycol fatty acid ester	N	4.5
Span 60	1	Sorbitan monostearate	N	4.7
Arlacel 60	1	Sorbitan monostearate	N	4.7
Atlas G-2139	1	Diethylene glycol monooleate	N	4.7
Emcol DO 50	2	Diethylene glycol fatty acid ester	N	4.7
Atlas G 2146	1	Diethylene glycol monostearate	N	4.7
Emcol DS-50	2	Diethylene glycol fatty acid ester	N	4.7
Atlas G-1702	1	Polyoxyethylene sorbitol beeswax derivative	N	5
Emcol DP-50	2	Diethylene glycol fatty acid ester	N	5.1
Aldo 28	4	Glycerol monostearate (self-emulsifying)	A	5.5

Tegin	5	Glycerol monostearate (self-emulsifying)	A	5.5
Emcol DM-50	2	Diethylene glycol fatty acid ester	N	5.6
Atlas G-1725	1	Polyoxyethylene sorbitol beeswax derivative	N	6
Atlas G-2124	1	Diethylene glycol monolaurate (soap free)	N	6.1
Emcol DL-50	2	Diethylene glycol fatty acid ester	N	6.1
Glaurin	4	Diethylene glycol monolaurate (soap free)	N	6.5
Span 40	1	Sorbitan monopalmitate	N	6.7
Arlacel 40	1	Sorbitan monopalmitate	N	6.7
Atlas G-2242	1	Polyoxyethylene dioleate	N	7.5
Atlas G-2147	1	Tetraethylene glycol monostearate	N	7.7
Atlas G-2140	1	Tetraethylene glycol monooleate	N	7.7
Atlas G-2800	1	Polyoxypropylene mannitol dioleate	N	8
Atlas G-1493	1	Polyoxyethylene sorbitol lanolin oleate derivative	N	8

TABLE 2.7 (Continued)

Name	Manufacturer[a]	Chemical Designation	Type[b]	HLB[c]
Atlas G-1425	1	Polyoxyethylene sorbitol lanolin derivative	N	8
Atlas G-3608	1	Polyoxypropylene stearate	N	8
Span 20	1	Sorbitan monolaurate	N	8.6
Arlacel 20	1	Sorbitan monolaurate	N	8.6
Emulphor VN-430	3	Polyoxyethylene fatty acid	N	9
Atlas G-1734	1	Polyoxyethylene sorbitol beeswax derivative	N	9
Atlas G-2111	1	Polyoxyethylene oxypropylene oleate	N	9
Atlas G-2125	1	Tetraethylene glycol monolaurate	N	9.4
Brij 30	1	Polyoxyethylene lauryl ether	N	9.5
Tween 61	1	Polyoxyethylene sorbitan monostearate	N	9.6
Atlas G-2154	1	Hexaethylene glycol monostearate	N	9.6
Tween 81	1	Polyoxyethylene sorbitan monooleate	N	10.0

Atlas G-1218	1	Polyoxyethylene esters of mixed fatty and resin acids	N	10.2
Atlas G-3806	1	Polyoxyethylene cetyl ether	N	10.3
Tween 65	1	Polyoxyethylene sorbitan tristearate	N	10.5
Atlas G-3705	1	Polyoxyethylene lauryl ether	N	10.8
Tween 85	1	Polyoxyethylene sorbitan trioleate	N	11
Atlas G-2116	1	Polyoxyethylene oxypropylene oleate	N	11
Atlas G-1790	1	Polyoxyethylene lanolin derivative	N	11
Atlas G-2142	1	Polyoxyethylene monooleate	N	11.1
Myrj 45	1	Polyoxyethylene monostearate	N	11.1
Atlas G-2141	1	Polyoxyethylene monooleate	N	11.4
PEG 400 monooleate	6	Polyoxyethylene monooleate	N	11.4
PEG 400 monooleate	7	Polyoxyethylene monooleate	N	11.4
Atlas G-2076	1	Polyoxyethylene monopalmitate	N	11.6
S-541	4	Polyoxyethylene monostearate	N	11.6

TABLE 2.7 (Continued)

Name	Manufacturer[a]	Chemical Designation	Type[b]	HLB[c]
PEG 400 monostearate	6	Polyoxyethylene monostearate	N	11.6
PEG 400 monostearate	7	Polyoxyethylene monostearate	N	11.6
Atlas G-3300	1	Alkyl aryl sulfonate	A	11.7
		Triethanolamine oleate	A	12
Atlas G-2127	1	Polyoxyethylene monolaurate	N	12.8
Igepal CA-630	3	Polyoxyethylene alkyl phenol	N	12.8
Atlas G-1431	1	Polyoxyethylene sorbitol lanolin derivative	N	13
Atlas G-1690	1	Polyoxyethylene alkyl aryl ether	N	13
S-307	4	Polyoxyethylene monolaurate	N	13.1
PEG 400 monolaurate	6	Polyoxyethylene monolaurate	N	13.1
Atlas G-2133	1	Polyoxyethylene lauryl ether	N	13.1
Atlas G-1794	1	Polyoxyethylene castor oil	N	13.3
Emulphor EL-719	3	Polyoxyethylene vegetable oil	N	13.3
Tween 21	1	Polyoxyethylene sorbitan monolaurate	N	13.3
Renex 20	1	Polyoxyethylene esters of mixed fatty and resin acids	N	13.5

Atlas G-1441	1	Polyoxyethylene sorbitol lanolin derivative	N	14
Atlas G-7596J	1	Polyoxyethylene sorbitan monolaurate	N	14.9
Tween 60	1	Polyoxyethylene sorbitan monostearate	N	14.9
Tween 80	1	Polyoxyethylene sorbitan monooleate	N	15
Myrj 49	1	Polyoxyethylene monostearate	N	15.0
Atlas G-2144	1	Polyoxyethylene monooleate	N	15.1
Atlas G-3915	1	Polyoxyethylene oleyl ether	N	15.3
Atlas G-3720	1	Polyoxyethylene stearyl alcohol	N	15.3
Atlas G-3920	1	Polyoxyethylene oleyl alcohol	N	15.4
Emulphor ON-870	3	Polyoxyethylene fatty alcohol	N	15.4
Atlas G-2079	1	Polyoxyethylene glycol monopalmitate	N	15.5
Tween 40	1	Polyoxyethylene sorbitan monopalmitate	N	15.6
Atlas G-3820	1	Polyoxyethylene cetyl alcohol	N	15.7
Atlas G-2162	1	Polyoxyethylene oxypropylene stearate	N	15.7
Atlas G-1471	1	Polyoxyethylene sorbitol lanolin derivative	N	16

TABLE 2.7 (Continued)

Name	Manufacturer[a]	Chemical Designation	Type[b]	HLB[c]
Myrj 51	1	Polyoxyethylene monostearate	N	16.0
Atlas G-7596P	1	Polyoxyethylene sorbitan monolaurate	N	16.3
Atlas G-2129	1	Polyoxyethylene monolaurate	N	16.3
Atlas G-3930	1	Polyoxyethylene oleyl ether	N	16.6
Tween 20	1	Polyoxyethylene sorbitan monolaurate	N	16.7
Brij 35	1	Polyoxyethylene lauryl ether	N	16.9
Myrj 52	1	Polyoxyethylene monostearate	N	16.9
Myrj 53	1	Polyoxyethylene monostearate	N	17.9
		Sodium oleate	A	18
Atlas G-2159	1	Polyoxyethylene monostearate	N	18.8
		Potassium oleate	A	20
Atlas G-263	1	*N*-cetyl *N*-ethyl morpholinium ethosulfate	C	25–30
		Pure sodium lauryl sulfate	A	~40

[a] 1 = Atlas Chemical Industries, Inc.; 2 = Emulsol Corporation; 3 = General Amline and Film Corporation; 4 = Glyco Products Company, Inc.; 5 = Goldschmidt Chemical Corporation; 6 = Kessler Chemical Company, Inc.; 7 = W. C. Hardesty Company, Inc.

[b] A = anionic; C = cationic; N = nonionic.

[c] HLB values, either calculated or determined, believed to be correct to number 1.

2.3.2 Correlation between the HLB Number and Other Properties

The HLB number as defined in the previous section is certainly only an empirically defined quantity. However, ever since the quantity was first defined by Griffin (2), a number of attempts have been made to give it theoretical significance by relating it to other surface properties of the emulsifier, or, ultimately to the structure of the surface active molecule (28). There are a number of articles in which the relation between the HLB number and the other properties are described (29–40).

Davies (41) has treated the HLB number as derived from a summation of structural factors as something akin to, say, the parachor. From this point of view, he attempted to resolve the structure of the emulsifier into component groups, each of which makes a contribution (negative or positive) to the total HLB number.

The group numbers obtained from several known structures of defined HLB are detailed in Table 2.8. For a given structure, the HLB number is calculated by substituting the group numbers into the relation:

$$\text{HLB} = 7 + (\text{hydrophilic group numbers}) - (\text{lipophilic group numbers}, \qquad (12)$$

where the last term on the right is usually $0.475n$, where n is the number of $—CH_2—$ groups in the lipophile. It should be noted that the $—CH_2—$ groups of the polyoxyethylene chain are not included in this total, since each ethylene oxide group is included in the count as a unit.

The results of the calculation of some HLB numbers from these group numbers are shown in Table 2.9. The adjustment is well made.

TABLE 2.8 HLB Group Numbers (41)

Hydrophilic Group Numbers	
—SO_4Na	38.7
—CO_2K	21.1
—CO_2Na	19.1
—N (tertiary amine)	9.4
Ester (sorbitan ring)	6.8
Ester (free)	2.4
—CO_2H	2.1
—OH (free)	1.9
—O—	1.3
—OH (sorbitan ring)	0.5
Lipophilic Group Numbers	
—CH —CH_2— CH_3— —CH—	−0.475
Derived Group Numbers	
—(CH_2—CH_2—O)—	0.33
—(CH_2—CH_2—CH_2—O)—	−0.15

2.3.3 Applications of the HLB Method

Since the HLB of a surfactant at different oil/water interfaces differs, a correction for the type of oil used is necessary in the HLB number method, which is known as the "required HLB numbers of oils" and these are given in Tables 2.10 and 2.11.

The assumption has always been made that HLB numbers were given by the weight average of HLB numbers of respec-

TABLE 2.9 A Comparison Between Experimental (Literature) and Calculated HLB Numbers (41)

Name	Literature (3)	Calculated
Tween 80	15	15.8
Tween 81	10	10.9
Span	8.6	8.5
Span 40	6.7	7.0
Span 60	4.7	5.7
Span 80	4.3	5.0
Glycerol monostearate	3.8	3.7
Span 65	2.1	2.1

TABLE 2.10 HLB Numbers for Various Emulsion Applications (3,24)

Application	Emulsion Type	HLB Range
Cream, all-purpose	O/W	6–8
Cream, antiperspirant	O/W	14–17
Cream, cold	O/W	7–15
Cream, stearic acid	O/W	6–15
Creams and lotions	W/O	4–6
Lotions	O/W	6–18
Oil, perfume	O/W	9–16
Oil, mineral	O/W	9–12
Oil, vegetable	O/W	7–12
Oil, vitamin	O/W	5–10
Ointment bases		
Absorption	W/O	2–4
Washable	O/W	10–12
Ointment, emollient	O/W	8–14
Polishes	O/W	8–12

TABLE 2.11 HLB Numbers Required to Emulsify Various Oil Phases (3,24)

Oil Phase	W/O Emulsion	O/W Emulsion
Acetophenone	—	14
Acid, dimer	—	14
Acid, lauric	—	16
Acid, linoleic	—	16
Acid, oleic	—	17
Acid, ricinoleic	—	16
Acid, stearic	—	17
Alcohol, cetyl	—	15
Alcohol, decyl	—	14
Alcohol, lauryl	—	14
Alcohol, tridecyl	—	14
Benzene	—	15
Carbon tetrachloride	—	16
Castor oil	—	14
Chlorinated paraffin	—	8
Cyclohexane	—	15
Kerosene	—	14
Lanolin, anhydrous	8	12
Oil		
Mineral, aromatic	4	12
Mineral, paraffinic	4	10
Mineral spirits	—	14
Petrolatum	4	7–8
Pine Oil	—	16
Propene, tetramer	—	14
Toluene	—	15
Wax		
Beeswax	5	9
Candelilla	—	14–15
Carnauba	—	12
Microcrystalline	—	10
Paraffin	4	10
Xylene	—	14

tive surfactants. It is now known that this is not strictly the case, for example, Becher and Birkmeier (42) have shown, by their gas–liquid chromatographic technique, that the retention-time ratios for mixtures is not linear with composition.

A more direct way of studying the HLB of surfactant blends is to determine the PIT (HLB temperature) of surfactant blends as a function of composition and to compare the PIT of a series of surfactants whose HLB numbers change from a hydrophilic to lipophilic surfactant, such as Tween 80 to Span 80 (43). It is known from such studies that hydrophilic emulsifiers and lipophilic emulsifier blends may separately dissolve in water and oil phases and the actual HLB number of the blend deviates significantly from the weight average. Such behavior can only be detected from the study of the emulsion itself, such as PIT (HLB temperature) measurements (43). If the lipophilic emulsifier is very soluble or the critical micellization concentration (cmc) in the oil phase is fairly high, the HLB of the adsorbed mixed monolayer of emulsifiers could have quite a different composition from the stoichiometric composition. Such effects cannot be reflected in the gas–liquid chromatographic retention-time ratio. The deviation of the actual HLB number from the weight average of the blend did not appear to be significant, but the deviation seems quite large (43). Too small of a required HLB number for the W/O-type emulsions may result from such effects.

Once a correct "required HLB number" is determined, we can say that the stability of the emulsion made with a given pair of emulsifiers as a function of composition or a series of emulsifiers as a function of the ethylene oxide chain length is most stable at the "required HLB number." The stability of the emulsion is also a function of the molecular size of the emulsifiers, interactions among surfactants, oil and water, molecular shape, and so on.

Thus, in formulating a given emulsion, one may take a pair of emulsifying agents not too widely different in their HLB numbers (about 5) and vary their net HLB numbers over the range in which they would be expected to be effective. Having found the most effective HLB number, one would then try various agents or a pair of agents until the most effective one was found.

The concept of the variation in the HLB number casts an illuminating sidelight on an old and well-known empirical rule. It has long been known that the efficiency of sodium oleate is increased by the presence of a small amount of free oleic acid. Sodium oleate has an HLB number of 18. The addition of even a small amount of oleic acid with its extremely low HLB (~ 1), will then give an HLB number in the middle of the suitable range.

2.4 HYDROPHILE–LIPOPHILE BALANCE, HLB, OF IONIC SURFACTANTS

It is known that a nonionic surfactant whose HLB is optimum for a particular application, is much more efficient than the other nonionic surfactant of different HLB or of different oxyethylene chain length. Now it is possible to change the HLB of ionic surfactants by changing the number of valencies and concentration of counterions and by replacing an ionic surfactant with lipophilic cosurfactant such as a fatty acid, alcohol, or polyoxyethylene glycol alkylether (44,45). With addition or replacement of a cosurfactant solubilization area increases remarkably (44–48), and the stability of the emulsion also increases. It is known also that an emulsion is stabilized by a condensed mixed film of cholesterol and sodium hexadecyl sulfate by the addition of cholesterol to an aqueous solution of $R_{16}SO_4Na$. Addition of multivalent ions depresses the surface charge at the interface and the HLB of the adsorbed film decreases. Finally inversion occurs in the emulsion (49). All these processes could be interpreted as the change of the HLB of adsorbed surfactant at the interface.

The evaluation of the HLB composition of an ionic surfactant may be performed by studying the phase diagram of a four component system composed of ionic surfactant, cosurfactant, water, and an oil system (cf. Figs. 1.17–1.20). (48)

As shown on the water axis in Fig. 1.17, the solution behavior of ionic surfactant + cosurfactant changes with the composition. Mixed surfactant dissolved in water forming micelles in an ionic surfactant rich region. If the ratio of cosurfactant increases, the liquid crystalline phase or surfactant phase separates. Namely, the dissolution behavior of ionic surfactant + cosurfactant in water may be used to estimate the HLB of ionic surfactant + cosurfactant mixture.

Ordinary ionic surfactants are usually too hydrophilic. However, bivalent salts of anionic surfactants such as $C_{12}H_{25}OCH_2CH_2OSO_3Mg^{1/2}$, (or $Ca^{1/2}$) (44), and ionic surfactants that have two comparable hydrocarbon chains (17,50) are well balanced surfactants and the solution behavior of sodium dialkyl carboxylates in water actually showed such a tendency as shown in Fig. 2.4.

Other Methods of Estimating the HLB of Ionic Surfactant

Lin and Marszall proposed an equation that relates the cmc and the HLB number of ionic surfactants (51). However, HLB means the balance of molecules. The cmc will decrease

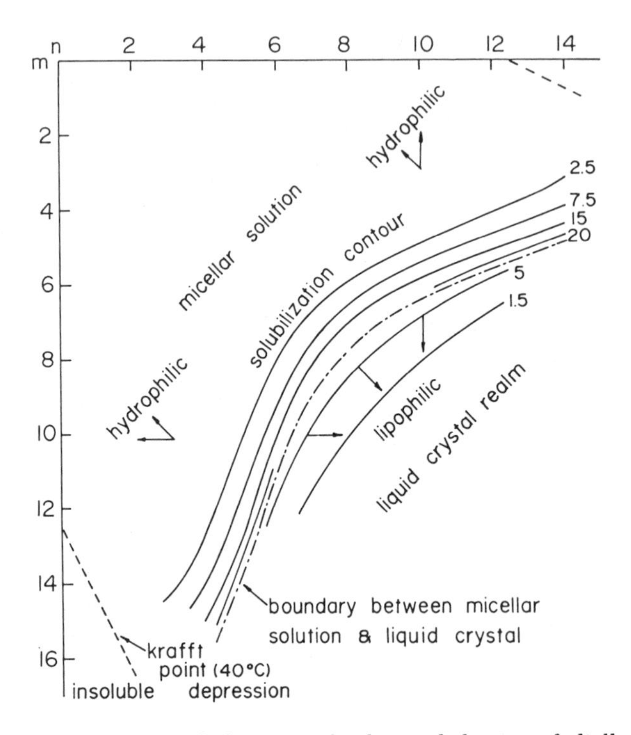

Figure 2.4 Conceptual diagram of solution behavior of dialkyl-type ionic surfactants

$$\left.\begin{matrix} C_mH_{2m+1} \\ C_nH_{2n+1} \end{matrix}\right\rangle CHCO_2Na$$

at 40°C. Solubilization contours were drawn based on the solubilization of C_6H_{12} g/100 g of solution in 0.164 mole/kg of aqueous surfactant solutions. [Reproduced by permission of Am. Chem. Soc., *J. Phys. Chem.*, 87, 2018 (1983).]

geometrically with the size of the hydrophile and lipophile groups while the HLB number is kept constant.

REFERENCES

1. W. Clayton, *Theory of Emulsions*, 4th ed., The Blakiston Co., Philadelphia (1943), p. 127.
2. W. C. Griffin, *J. Soc. Cosmet. Chem.*, **1**, 311 (1949).
3. W. C. Griffin, *J. Soc. Cosmet. Chem.*, **5**, 249 (1954).
4. J. T. Davies, *Interfacial Phenomena*, Academic, New York (1963), p. 371–383.
5. C. D. Moore and M. Bell, *Soap, Perfum. Cosmet.*, **29**, 893 (1956).
6. H. L. Greenwald, G. L. Brown, and M. N. Fineman, *Anal. Chem.*, **28**, 1693 (1956).
7. K. Shinoda and H. Arai, *J. Phys. Chem.*, **68**, 3485 (1964).
8. K. Shinoda, *J. Colloid Interface Sci.*, **24**, 4 (1967).
9. K. Shinoda and H. Saito, *J. Colloid Interface Sci.*, **30**, 258 (1969).
10. K. Shinoda, Proc. 5th Int. Congr. Surface Active Substances, Barcelona, 3, 275 (1968).
11. K. Shinoda and H. Sagitani, *J. Colloid Interface Sci.*, **64**, 68 (1978).
12. L. Marszall, *Acta Pol. Pharm.*, **32**, 397 (1975); L. Marszall, *Cosmet. Toiletries*, **90**, 37 (1975); **92**, 32 (1977).
13. L. Marszall, *Colloid Polymer Sci.*, **254**, 674 (1976).
14. L. Marszall, *Fette, Seifen Anstrichm.*, **80**, 289 (1978).
15. K. Shinoda, H. Kunieda, N. Obi, and S. Friberg, *J. Colloid Interface Sci.*, **80**, 304 (1981).
16. Section 1.3 in Chapter 1.

17. K. Shinoda and H. Sagitani, *J. Phys. Chem.*, **87**, 2018 (1983).
18. H. Saito and K. Shinoda, *J. Colloid Interface Sci.*, **24**, 10 (1967).
19. K. Shinoda and T. Ogawa, *J. Colloid Interface Sci.*, **24**, 56 (1967).
20. K. Shinoda and H. Saito, *J. Colloid Interface Sci.*, **26**, 70 (1968).
21. S. E. Friberg and I. Lapczynska, *Prog. Colloid Polymer Sci.*, **55**, 614 (1976).
22. S. E. Friberg, I. Lapczynska, and G. Gillberg, *J. Colloid Interface Sci.*, **56**, 19 (1976).
23. H. Kunieda and K. Shinoda, *J. Colloid Interface Sci.*, **75**, 601 (1980).
24. Atlas Chemical Industries, Inc., *The Atlas HLB System*, 2nd ed., (revised) Wilmington, Delaware (1963).
25. K. Shinoda, H. Saito, and H. Arai, *J. Colloid Interface Sci.*, **35**, 624 (1971).
26. P. Becher, *Emulsions: Theory and Practice*, 2nd ed., Reinhold, London (1965), pp. 231–235.
27. W. C. Griffin, *Off. Dig. Fed. Paint Varn. Prod. Clubs*, **28**, 466 (1956).
28. P. Becher, *J. Soc. Cosmet. Chem.*, **76**, 33 (1961).
29. V. W. Wachs and S. Hayano, *Kolloid Z.*, **181**, 139 (1962).
30. S. Hayano and T. Asahara, Proc. 5th Int. Congr. Surface Active Substances, Barcelona, **2** (Part 2), 843 (1968).
31. L. Marszall, *Acta Pol. Pharm.*, **31**, 671 (1974).
32. R. Heusch, *Kolloid Z.*, **236**, 1 (1970); *Fette Seifen Anstrichm.*, **72**, 969 (1970).
33. I. Racz and E. Orban, *J. Colloid Sci.*, **20**, 99 (1965); E. Orban, *Tenside Deterg.*, **7**, 203 (1970).
34. L. Marszall, *Kolloid Z.*, **251**, 609 (1973).

35. H. Schott, *J. Pharm. Sci.*, **60**, 648 (1971).
36. P. M. Kruglyakov and A. F. Koretzky, *Dokl. Akad. Nauk SSSR*, **197**, 1106 (1971).
37. C. McDonald, *Can. J. Pharm. Sci.*, **5**, 81 (1970).
38. L. Marszall, *J. Pharm. Pharmacol.*, **25**, 254 (1973).
39. W. G. Gorman and G. D. Hall, *J. Pharm. Sci.*, **52**, 442 (1963).
40. N. Ohba, *Bull. Chem. Soc. Jpn.*, **35**, 1016 (1962).
41. J. T. Davies, *Proc. Int. Congr. Surface Activity*, 2nd, London, **1**, 426 (1957).
42. P. Becher and R. L. Birkmeier, *J. Am. Oil Chem. Soc.*, **41**, 169 (1964).
43. K. Shinoda, T. Yoneyama, and H. Tsutsumi, *J. Dispersion Sci. Technol.*, **1**, 1 (1980).
44. K. Shinoda and T. Hirai, *J. Phys. Chem.*, **81**, 1842 (1977).
45. K. Shinoda, *Pure Appl. Chem.*, **52**, 1195 (1980).
46. H. B. Klevens, *J. Chem. Phys.*, **17**, 1004 (1949).
47. H. B. Klevens, *J. Am. Chem. Soc.*, **72**, 3581, 3780 (1950).
48. K. Shinoda, H. Kunieda, T. Arai, and H. Saijo, *J. Phys. Chem.*, 88, 5126 (1984)
49. F. Z. Saleeb, C. J. Canto, T. K. Streckfus, J. R. Frost, and H. L. Rosano, *J. Am. Oil Chem. Soc.*, **52**, 208 (1975).
50. H. Kunieda and K. Shinoda, *J. Phys. Chem.*, **82**, 1710 (1977).
51. I. J. Lin and L. Marszall, *J. Colloid Interface Sci.*, **57**, 85 (1976).

CHAPTER 3

Factors Affecting the Phase Inversion Temperature (PIT) in an Emulsion

Various factors such as the oxyethylene chain length of the emulsifier, size of the hydrophilic group, the types of oil, additives in water and in oil, phase volume, mixing of emulsifiers, and so on, on the phase inversion temperature (PIT) will be presented and their relative importance discussed. The change of PIT (HLB temperature) in an emulsion means a change of the emulsifier hydrophile–lipophile balance, HLB, at the oil/water interface.

3.1 INFLUENCE OF THE TYPES OF OILS ON THE PIT

The PIT versus the emulsifier concentration curves for various types of hydrocarbons are shown in Figs. 3.1 and 3.2 (1,2).

The PIT in an emulsion differs widely for different oils. The relation to the solubility of the emulsifier in the various hydrocarbons is straightforward. The PIT in benzene/water is the lowest of any of the hydrocarbons examined. Hydrated polyoxyethylene(9.6)nonylphenylether dissolves well in benzene above ≈20°C. On the other hand, hexadecane or liquid paraffin, which is the poorest solvent for the same emulsifier, gave the highest PIT, about 110°C. Hence, the more soluble the oil for a definite hydrated nonionic emulsifier, the lower was the PIT. The higher the cloud point in aqueous surfactant solution saturated with various oils, the higher was the PIT. This rule (1) coincides with Bancroft's rule that the phase in which the emulsifying agent is more soluble will be the external one at a definite temperature (3).

The cloud points showed a similar relationship when the influence of the hydrocarbon was taken into consideration.

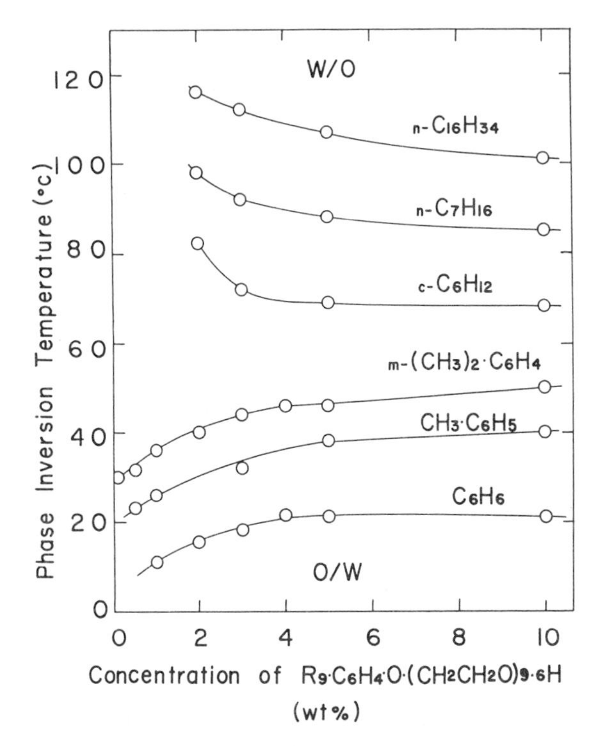

Figure 3.1 The effect of different hydrocarbons on the PIT of emulsions (volume ratio = 1) versus the concentration of polyoxyethylene(9.6) nonylphenylether (wt% in water). [Reproduced by permission of *Am. Chem. Soc.*, *J. Phys. Chem.*, **68**, 3485 (1964).]

The cloud point of polyoxyethylene(9.6)nonylphenylether is approximately 63°C, but the relevant value is the one in the presence of solubilized oils. The cloud points in aqueous surfactant solutions saturated with various hydrocarbons and the PIT in emulsions composed of respective hydrocarbons and water are close (4).

The PIT versus the phase volume curves for various types of hydrocarbons, halogenocarbons, and polymethylphenyl-

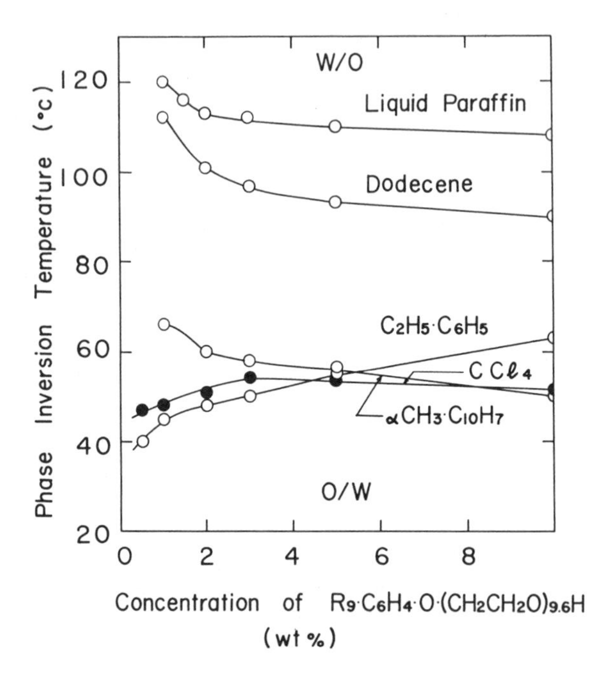

Figure 3.2 The effect of different hydrocarbons on the PIT versus the concentration of polyoxyethylene(9.6)nonylphenylether. [Reproduced by permission of *Am. Chem. Soc.*, *J. Phys. Chem.*, **68**, 3485 (1964).]

siloxane are shown in Figs. 3.3 (5) and 3.4 (6). The PITs are fairly constant over a wide volume fraction range for these (relatively) nonpolar and saturated oils, and indicate a rather strong emulsion-type-determining influence of nonionic emulsifiers. From the PITs of respective oils for a given surfactant, we can estimate the order of oils that require more hydrophilic or more lipophilic emulsifier to obtain O/W- or W/O-type emulsions.

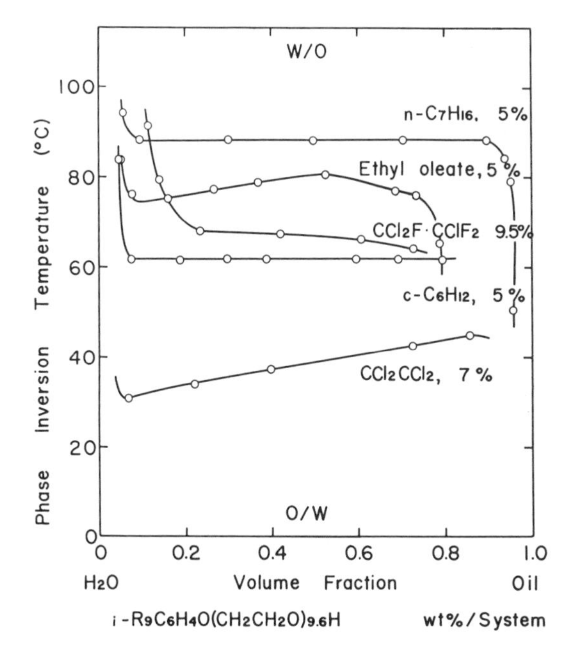

Figure 3.3 The effect of the phase volume on the PIT of emulsions stabilized with polyoxyethylene(9.6)nonylphenylether. The types of oils and the concentration of surfactant (wt%) are indicated in the graph. [Reproduced by permission of Academic, *J. Colloid Interface Sci.*, 25, 429 (1967).]

3.2 EFFECT OF THE OXYETHYLENE CHAIN LENGTH OF EMULSIFIER ON THE PIT

It is expected that the longer the hydrophilic chain, the higher the cloud point and/or the PIT would be. This relation has been determined in benzene/water and hexadecane/water systems for a homologous series of polyoxyethylene nonylphenylethers and is shown in Figs. 3.5 and 3.6 (1).

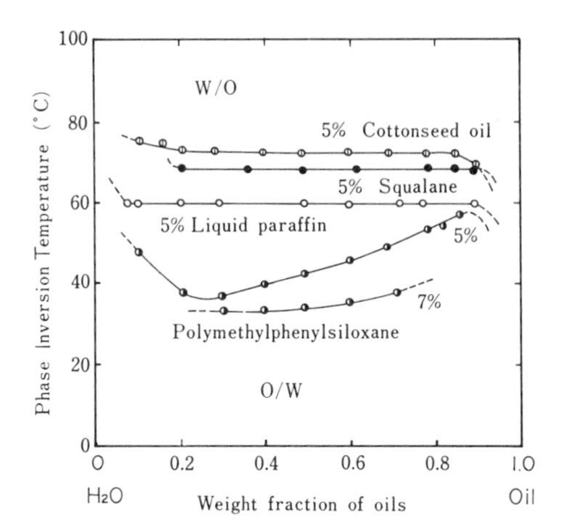

Figure 3.4 The effect of the weight fraction of oils on the PIT of emulsion stabilized with $C_9H_{19}C_6H_4O(CH_2CH_2O)_{5.8}H$. The types of oils and the concentration of surfactant (wt%) are indicated in the graph.

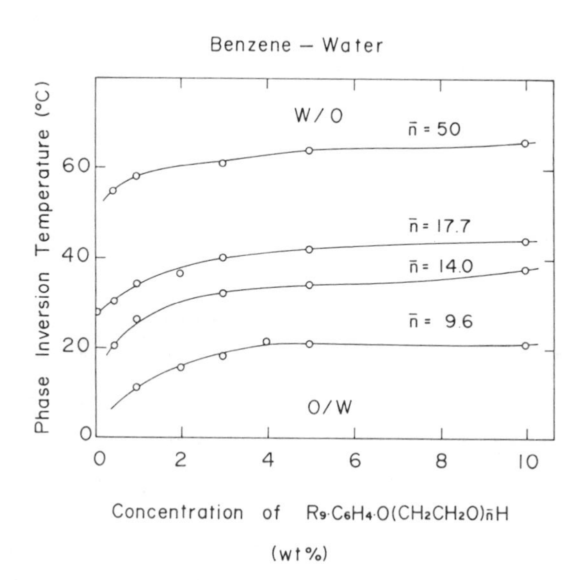

Figure 3.5 The effect of the hydrophilic chain length of the emulsifier on the PIT in benzene/water emulsions (volume ratio = 1). [Reproduced by permission of *Am. Chem. Soc., J. Phys. Chem.*, **68**, 3485 (1964).]

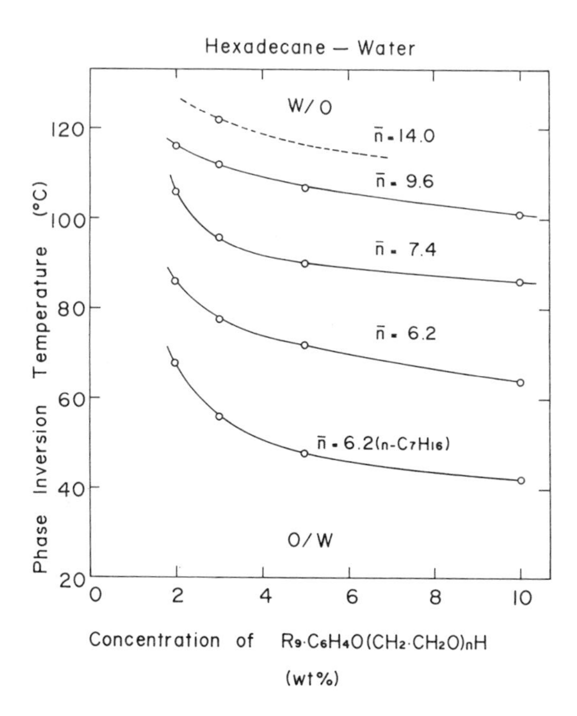

Figure 3.6 The effect of the hydrophilic chain length of the emulsifier on the PIT in hexadecane/water emulsions (volume ratio = 1). [Reproduced by permission of *Am. Chem. Soc., J. Phys. Chem.*, **68**, 3485 (1964).]

The PIT stays almost constant at higher concentration ranges, but it is raised at lower concentrations when the oil is a saturated hydrocarbon. This occurs because shorter oxyethylene chain homologs of surfactant preferentially dissolve in the oil phase and the average oxyethylene chain length of surfactants at the oil/water interface becomes longer, that is, more hydrophilic. On the contrary, aromatic hydrocarbons dissolve well in the oxyethylene portion of the surfactant so that the PIT is depressed. Longer oxyethylene chain homologs of surfactant will also dissolve well in aromatic hydrocarbons whose PIT does not change much at lower surfactant concentration. Hence, a polyoxyethylene-type nonionic surfactant may not be a good surfactant for aromatic hydrocarbons.

The effect of the oxyethylene chain length of nonylphenylethers (emulsifiers) on the PITs in emulsions of various oil/water (1 : 1) weight systems are shown in Fig. 3.7. The longer the hydrophilic chain length of the surfactant, the higher the PIT.

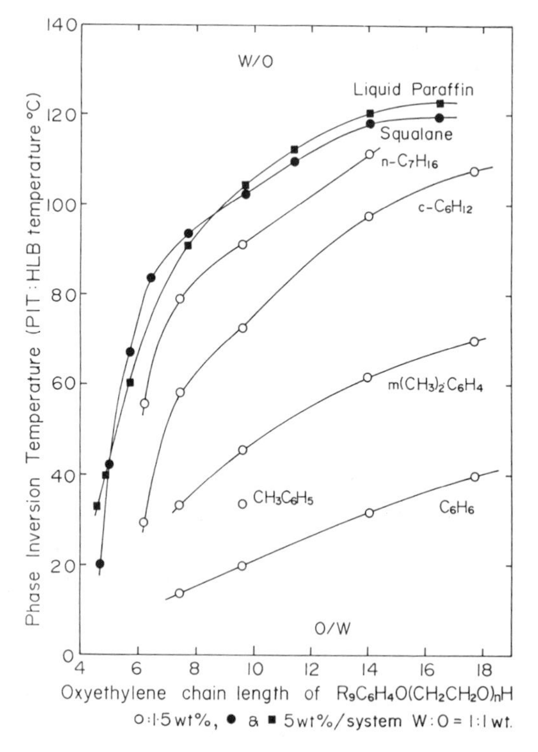

Figure 3.7 The correlation between the oxyethylene chain length of polyoxyethylene nonylphenylether and the PIT in emulsions of various oils.

3.3 THE EFFECT OF PHASE VOLUME ON THE PIT OF EMULSIONS STABILIZED WITH NONIONIC EMULSIFIERS

The PIT is considered to be a temperature at which the hydrophile–lipophile property of a surfactant just balances for a given system, so that an emulsion type inverts when the temperature passes the PIT. The HLB number of a homologous series of nonionic surfactants changes with the hydrophilic chain length of the emulsifier. This change is compensated for by the change of temperature, so the HLB number of respective emulsifiers at respective PITs is considered to be the same.

In contrast to the temperature that has a marked effect on the HLB and emulsion type, the effect of phase volume on the PIT is relatively small. Actually, the PIT is fairly constant over a wide volume fraction range in many nonpolar solvents, such as heptane, cyclohexane, and halogenocarbons, as shown in Figs. 3.3 and 3.4. This fact indicates the strong type-determining influence of nonionic surfactants in emulsions, in contrast to the influence of ionic agents (7,8). Phase volume is not very important, but temperature is a very important factor in controlling the types of emulsions stabilized with nonionic surfactants.

The effect of the hydrophilic chain length of the nonionic surfactant on the PIT versus the phase–volume curve of cyclohexane/water emulsions is shown in Fig. 3.8 (5). The shorter the hydrophilic chain length of an emulsifier, the lower the PIT.

The PIT versus phase–volume curve does not change strongly with the concentration of surfactant in concentrated solutions (above 5 wt%), but does change in dilute solution (below 3 wt%), as shown in Fig. 3.9. This trend is

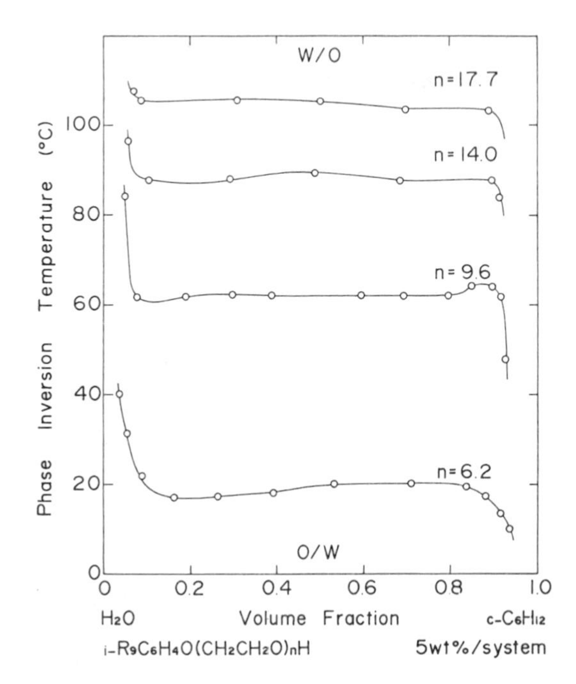

Figure 3.8 The effects of the hydrophilic chain length of nonionic surfactants on the PIT versus phase–volume curves of cyclohexane/water emulsions. [Reproduced by permission of Academic, *J. Colloid Interface Sci.*, **25**, 429 (1967).]

explained as follows: The saturation concentrations of nonionic homologues in water are all very small, but those in hydrocarbon are much larger and depend largely on the thyleneoxide chain length. Lipophilic homologues, that is, shorter oxyethylene chain compounds, dissolve better than hydrophilic homologues in an oil phase (9–12). Hence, there will be a selection of more hydrophile emulsifiers to be adsorbed at the oil/water interface. This effect is amplified in dilute solutions and in the case when the volume fraction of oil is large.

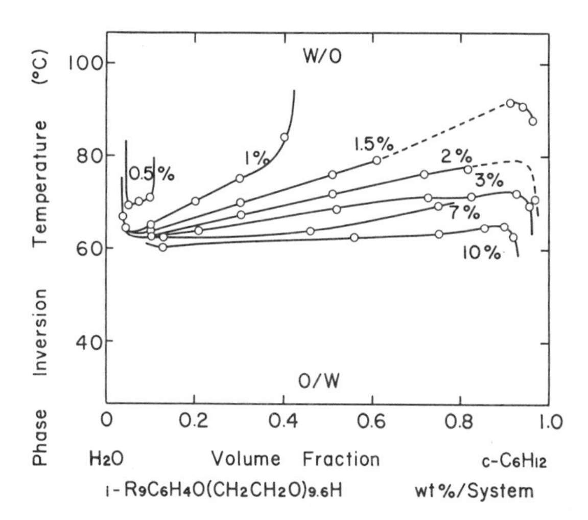

Figure 3.9 The effect of the concentration of surfactant on the PIT versus phase–volume curve. [Reproduced by permission of Academic, *J. Colloid Interface Sci.*, **25**, 429 (1967).]

Combining this information, a close correlation is found among the PIT in emulsions, the cloud point in aqueous solutions saturated with oil, and the cloud point in nonaqueous solutions saturated with water. Accordingly, in these systems, the phase diagram and the PIT curve can be estimated from the PIT data in emulsions of 1 : 1 volume ratio, provided the concentration of nonionic surfactant is reasonably high. Thus, the relation between the PIT in an emulsion (1 : 1 volume ratio) and the hydrophilic chain length of nonionic surfactants is very important.

The PIT versus phase–volume curves of cyclohexane/water emulsions stabilized with the surfactant mixtures (3 wt%) are shown in Fig. 3.10 (5).

Since the distribution of the oxyethylene chain length in

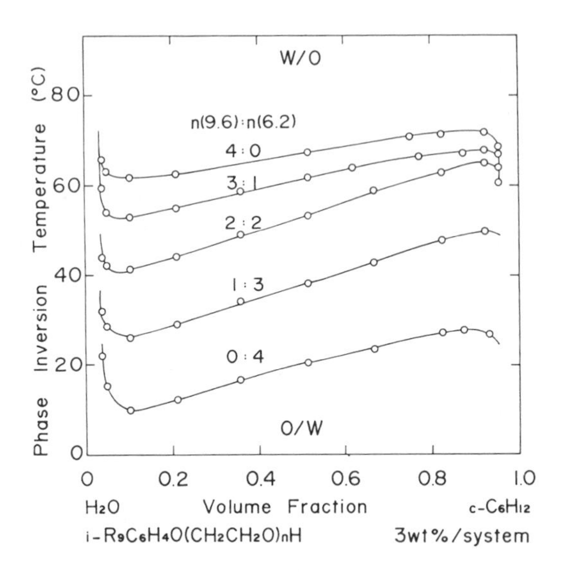

Figure 3.10 The PIT versus phase–volume curves of cyclohexane/water emulsions stabilized with the mixtures of polyoxyethylene(9.6)nonylphenylether (3 wt%). Mixing ratios (weight) are indicated in the graph. [Reproduced by permission of Academic, *J. Colloid Interface Sci.*, **25**, 429 (1967).]

mixed emulsifier is broader, the PIT is raised more rapidly with the volume fraction of oil.

On the other hand, the PIT depends on the phase volume in some other solvents such as benzene, bromobenzene (chlorobenzene), and *m*-xylene, as shown in Fig. 3.11. Dodecane behaves halfway between saturated hydrocarbons and aromatic oils. The PIT in these solvents is also affected by the way of shaking. (The present results were obtained by vertical shaking.) Hence, in emulsions composed of these oils, the predictions of the solubilization curve, the PIT, and the optimum temperature in emulsions are less accurate.

Thus, it is clear that the properties of emulsions of some

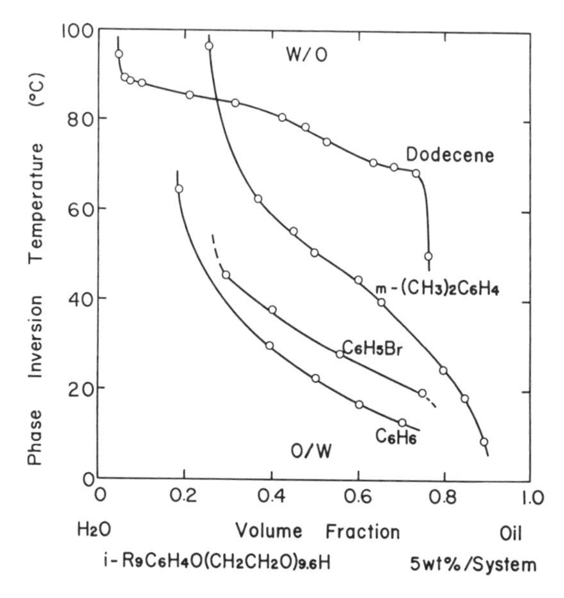

Figure 3.11 The effect of phase volume on the PIT of emulsions stabilized with polyoxyethylene(9.6)nonylphenylether (5 wt%). The types of oils are indicated in the graph. [Reproduced by permission of Academic, *J. Colloid Interface, Sci.*, **25**, 429 (1967).]

kinds of oils are intrinsically different from those of saturated hydrocarbons, and therefore all types of oils cannot be treated uniformly by simple HLB number or HLB temperature (PIT). Information of this sort, however, can be obtained from the study of the PIT curves and serves to classify various types of oils. Saturated hydrocarbons and halogenocarbons behave quite normally, as do unsaturated hydrocarbons and esters at high surfactant concentration. On the other hand, aromatic hydrocarbons and polar oils, such as benzene, chloroform, nitrobenzene, and alcohols, behave quite abnormally in regard to the behavior of emulsions, especially at low surfactant concentration.

3.4 THE EFFECT OF THE HYDROCARBON CHAIN LENGTH OF OILS ON THE PIT OF EMULSIONS

The effect of the hydrocarbon chain length of *n*-alkanes on the PIT is summarized in Fig. 3.12. The longer the chain length of the hydrocarbon, the higher the PIT.

3.5 THE PIT AS A FUNCTION OF THE COMPOSITION OF OIL MIXTURES

The PITs of emulsions of oil/water mixtures (1 : 1 volume) as a function of the composition of oil are shown in Figs. 3.13 and 3.14 (13).

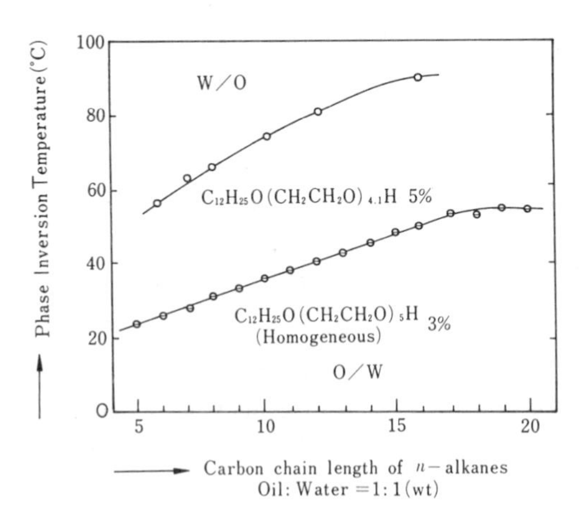

Figure 3.12 The correlation between the hydrocarbon chain length of *n*-alkanes and the PIT (HLB temperature).

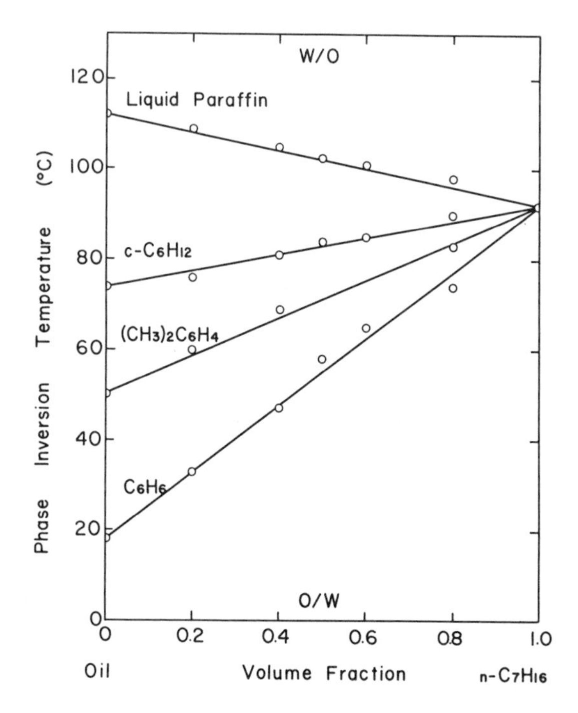

Figure 3.13 The effect of the mixing of n-heptane with various oils on the PITs of emulsions stabilized with polyoxyethylene(9.6)nonylphenylethers (1.5 wt%). [Reprinted by permission of Academic, New York *J. Colloid Interface Sci.*, **25**, 396 (1967)].

The PITs of emulsions in which the oil phase consists of oil mixtures is expressed by the volume average of the PITs of the respective oils, provided the two types of oils are similar.

$$\text{PIT (mixture)} = \text{PIT (1)}\phi_1 + \text{PIT (2)}\phi_2 \qquad (3.1)$$

where ϕ_1 and ϕ_2 are the volume fractions of oils 1 and 2, and the PIT (mixture), PIT (1) and PIT (2), are the PIT of emulsions of the oil mixture, oil 1 and oil 2, respectively. This con-

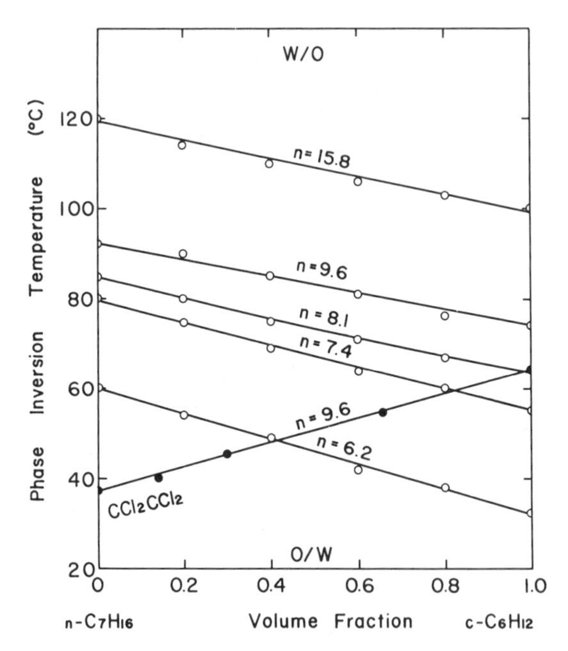

Figure 3.14 The effect of the mixing of *n*-heptane and cyclohexane on the PITs of emulsions stabilized with polyoxyethylene(n)nonylphenylethers (3 wt%). Oxyethylene chain lengths are indicated in the graph. Filled circles express the effect of the mixing (weight fraction) of cyclohexane and perchloroethylene on the PITs with polyoxyethylene(9.6)nonylphenylether (7 wt%). [Reproduced by permission of Academic, *J. Colloid Interface Sci.*, **25**, 396 (1967).]

clusion agrees with the empirical relation that the required HLB number of oil mixtures is calculated as an arithmetic average of the respective HLB number of oils (14).

3.6 THE PIT OF EMULSIONS OF EMULSIFIER MIXTURES

The PIT in *n*-heptane/water and cyclohexane/water systems emulsified with mixtures of similar emulsifiers, for example,

$R_9C_6H_4O(CH_2CH_2O)_{15.8}H$ and $R_9C_6H_4O(CH_2CH_2O)_{7.4}H$, are shown in Fig. 3.15. The PIT changes monotonously between the PITs of the individual emulsifiers. The PITs of emulsions stabilized with the surfactant mixtures are always higher than would be predicted from the weight average of the PITs of the constituent surfactants. This result is understandable because the shorter oxyethylene chain emulsifier is more soluble in the oil phase and the HLB number at the interface is shifted to a higher value on the one hand and the correlation between the PIT (HLB temperature) and the

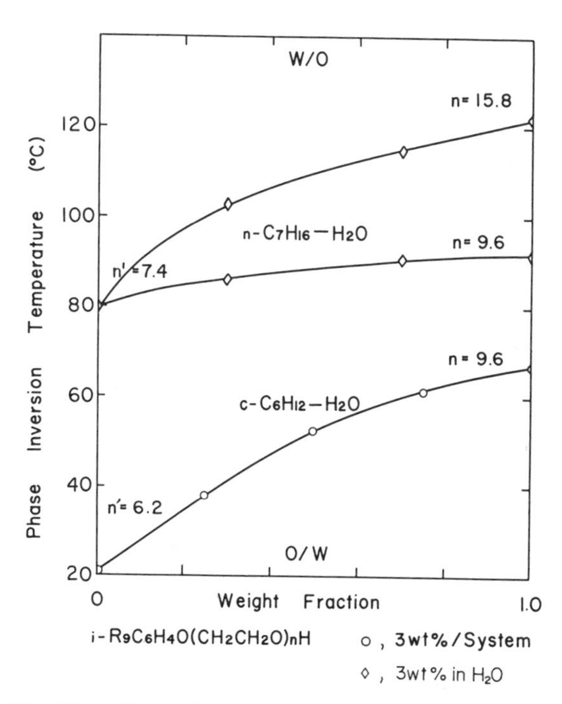

Figure 3.15 The effect of the mixing of surfactants on the PITs of *n*-heptane/water and cyclohexane/water emulsions. (◇ = 3 wt% in water) (○ = 3 wt%). [Reproduced by permission of Academic, *J. Colloid Interface Sci.*, **25**, 396 (1967).]

HLB number for various surfactants is not linear but convex as shown in Fig. 3.16. Since the HLB number of the emulsifier at the n-C_7H_{16}/H_2O interface is about 10 at the PIT, Fig. 3.16 is useful to find the change of the HLB number from 25°C to that at the respective PIT.

It is known also that the HLB number of the surfactant mixture does not strictly obey the weight average relationship (15–19).

Blends of a hydrophilic emulsifier, for example, Tween (eicosaoxyethylene sorbitan monoester) and a lipophilic emulsifier, for example, Span (sorbitan monoester) have been recommended because the HLB number of two emulsifiers are very different and it is said that any HLB number between these two emulsifiers is obtainable in blending these emulsifiers (14,20). We experience that the more stable emulsions are usually obtained by blending two or more emulsifiers rather than a single emulsifier whose HLB coincides with the required HLB (21–23), but the stability of the emulsion decreases when strongly hydrophilic and strongly lipophilic emulsifiers are mixed (22).

The PIT in emulsions of water/liquid paraffin (1:1) containing 2 wt% per system of polyoxyethylene(0,2,5, 7,10,15,20)sorbitan monooleates and their mixtures have been determined and plotted in Fig. 3.17 (24).

The HLB numbers of these emulsifiers were determined from their saponification number of the ester, S, and acid number of the acid, A, by the aid of Griffin's equation.

$$\text{HLB number} = 20(1 - S/A)$$

Analytical values agreed well with the assigned HLB numbers by Atlas. These values are summarized in Table 3.1. The HLB numbers of emulsifier blends have been calculated assuming weight average (14). It is evident from Fig.

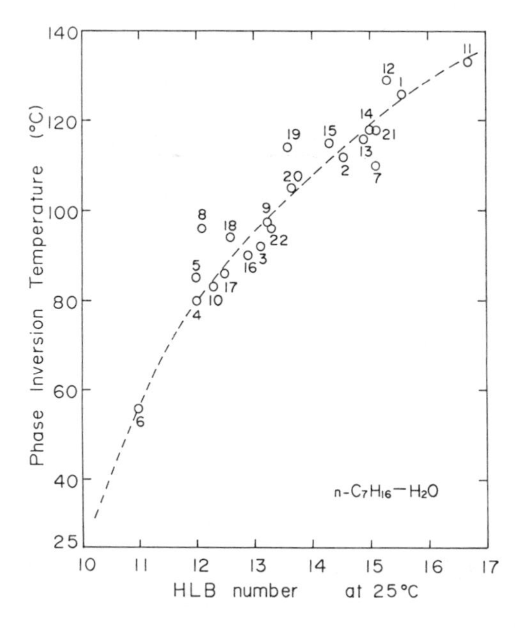

Figure 3.16 The correlation between the HLB number and the PIT (HLB temperature) of *n*-heptane/water emulsions stabilized with various surfactants (3 wt% for water). Weight ratios are used for mixtures. Key: 1, $R_9C_6H_4O(CH_2CH_2O)_{17.7}H$; 2, $R_9C_6H_4O(CH_2CH_2O)_{14.0}H$; 3, $R_9C_6H_4O(CH_2CH_2O)_{9.6}H$; 4, $R_9C_6H_4O(CH_2CH_2O)_{7.4}H$; 5, $R_{12}C_6H_4O$-$(CH_2CH_2O)_{9.0}H$; 6, $R_9C_6H_4O(CH_2CH_2O)_{6.2}H$; 7, $R_{12}O(CH_2CH_2O)_{13.0}H$; 8, $R_{12}O(CH_2CH_2O)_{6.5}H$; 9, Polyoxyethylene(13.9)styrenephenylether; 10, Polyoxyethylene(11.4)styrenephenylether; 11, Tween 20; 12, Tween 40; 13, Tween 60; 14, Tween 80; 15, $R_9C_6H_4O(CH_2CH_2O)_{15.8}H/R_9C_6H_4O(CH_2CH_2O)_{7.4}H$ = 7/3; 16, $R_9C_6H_4O(CH_2CH_2O)_{9.6}H/R_9C_6H_4O(CH_2CH_2O)_{7.4}H$ = 7/3; 17, $R_9C_6H_4O(CH_2CH_2O)_{9.6}H/R_9C_6H_4O(CH_2CH_2O)_{7.4}H$ = 3/7; 18, $R_{12}O(CH_2CH_2O)_{6.5}H/R_9C_6H_4O(CH_2CH_2O)_{9.6}H$ = 1; 19, $R_{12}O(CH_2CH_2O)_{6.5}H/R_9C_6H_4O(CH_2CH_2O)_{15.8}$ = 1; 20, $R_{12}O(CH_2CH_2O)_{13.0}H/R_9C_6H_5O(CH_2CH_2O)_{7.4}$ = 1; 21, $R_{12}O(CH_2CH_2O)_{13.0}H/R_9C_6H_4O(CH_2CH_2O)_{15.8}H$ = 1; 22, $R_9C_6H_5O(CH_2CH_2O)_{7.4}H$/Tween 80 = 1. [Reproduced by permission of Academic, *J. Colloid Interface Sci.*, **25**, 396 (1967).]

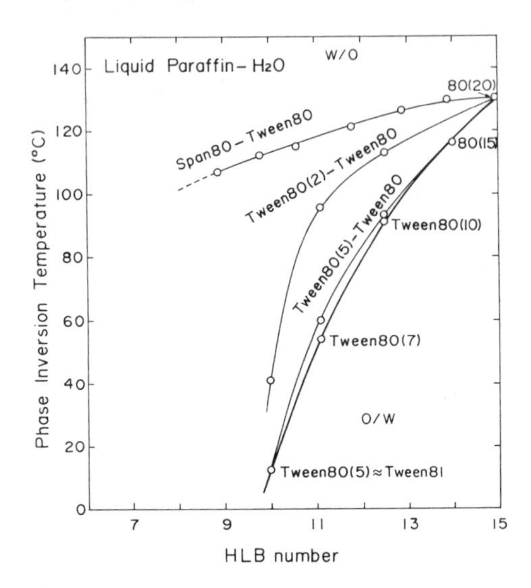

Figure 3.17 The PIT in emulsions composed of water/liquid paraffin (1 : 1 weight ratio) containing 2 wt% of polyoxyethylene(0,2,5, 7,10,15,20)sorbitan monooleates and their mixtures. [Reproduced by permission of Marcel Dekker, *J. Dispersion Sci. Tech.*, 1, 1 (1980).]

3.17 that the PIT versus HLB number curve of the mixtures of Tween 80(5) and Tween 80(20) is close to that of respective unmixed emulsifiers (broad curve). This means that emulsifiers very similar to polyoxyethylene(7,10,15)sorbitan monooleates are obtained by mixing an appropriate amount of Tween 80 and pentaoxyethylene sorbitan monooleate [Tween 80(5) = Tween 81]. The deviation is bigger in the mixtures of Tween 80(2) and Tween 80(20). Further deviation was observed in the mixture of Span 80 and Tween 80(20). A large amount of lipophilic emulsifier is necessary to depress the PIT or HLB of the emulsifier blend. This means the HLB of emulsifier blend does not change with the weight fraction of emulsifiers.

TABLE 3.1 The HLB Numbers and HLB Temperatures (PIT) of Tween-Type Emulsifiers (24)

Emulsifier	HLB Number by Atlas	HLB Number $20(1 - S/A)$	PIT (°C) (Liquid Paraffin)
Tween 80(20)[a]	15.0	15.0	130
Tween 80(15)	—	14.0	116.5
Tween 80(10)	—	12.5	91.3
Tween 80(7)	—	11.1	54.0
Tween 80(5) = Tween 81	10.0	10.0	12.3
Tween 80(2)	—	7.2	below 0
Span 80	4.3	4.4	below 0

[a]The number in brackets is the number of oxyethylene groups in a molecule. Tween 80(20) = Tween 80, Tween 80(5) = Tween 81, and Tween 80(0) = Span 80.

Deviation of the HLB numbers of emulsifier blends from the weight average is observed by Ohba (16) based on emulsion stability and by Becher and Birkmeier (19) based on their gas–liquid chromatographic technique. Becher said "All the data indicate that the deviation is rarely more than one or two HLB numbers and in many cases is much less." That is true when the difference in HLB numbers of emulsifiers is not large. But, the deviation from the PIT versus HLB number relation (broad curve) is more than three in the blends of Span 80 and Tween 80. For example, the PIT is 113°C when the calculated HLB number of the blend is 10, but the PIT = 113°C in this system corresponds to the HLB number = 13.7 as shown in Fig. 3.17. Such deviation observed in the figure is understandable, because the surface activity of the lipophilic emulsifier at the oil/water interface may be much small than that of the hydrophilic emulsifier. Furthermore, the PIT suddenly dropped below 0°C and could not be observed due to the further replacement of Tween 80 by Span 80. An emulsifier blend whose PIT ranges from 0–100°C was not obtained. This fact implies that emulsifier blends whose HLB number range is from 7–13 may not be prepared by blending Span 80 and Tween 80. A strongly hydrophilic emulsifier such as Tween may dissolve in the water phase, whereas an oleophilic surfactant such as Span 80 may dissolve mostly in the oil phase, so that the composition of emulsifiers in aqueous solution does not change with stoichiometric composition and vice versa the intermediate HLB cannot be prepared. Gas-chromatographic measurements cannot reflect such phenomena in emulsions.

Furthermore, Fig. 3.17 tells us that the blends of Tween 80(5)–Tween 80 and Tween 80(2)–Tween 80 can cover HLB numbers from 10–15 and the deviations from a weight average curve are about 0.2 and 1–1.5, respectively. Although the PIT is too low to be observed in the mixture of Span 80 and Tween 80(5), the HLB number of these emulsifier blends may change from 4.4–10 with the composition change, because the differences of the HLB numbers between Span 80 versus Tween 80(5) and Tween 80(5) versus Tween 80(20) are the same magnitude. Actually the use of a medium emulsifier, such as Tween 81–Tween 80(5), is usually known to yield more stable emulsions and is manufactured (24).

Figure 3.17 adds more evidence that the PIT (HLB temperature) study serves to clarify the understanding of the emulsification phenomena.

3.7 THE EFFECT OF ADDED SALTS, ACID, AND ALKALI ON THE PIT OF EMULSIONS

Although an HLB number is assigned to respective surfactants, the real HLB of surfactant at the oil/water interface changes with the amount and kinds of added salts in water as well as with the types of oils. The PIT of emulsions, on the other hand, accurately reflects the real HLB of surfactant in a given system. The effect of added salts, acid, and alkali in water on the hydrophile–lipophile balance of nonionic surfactant can be estimated from the measurements of the effect of these additives on the PIT of emulsions and on the phase diagram of the surfactant solutions.

The effect of added sodium chloride on the phase diagram of a water/cyclohexane system containing 5 wt% of polyoxyethylene(9.7)nonylphenylether as a function of temperature is shown in Fig. 3.18 (25). Points A, B, C, and D represent the cloud point in the absence of oil, the cloud point in aqueous surfactant solution saturated with oil, the PIT in the

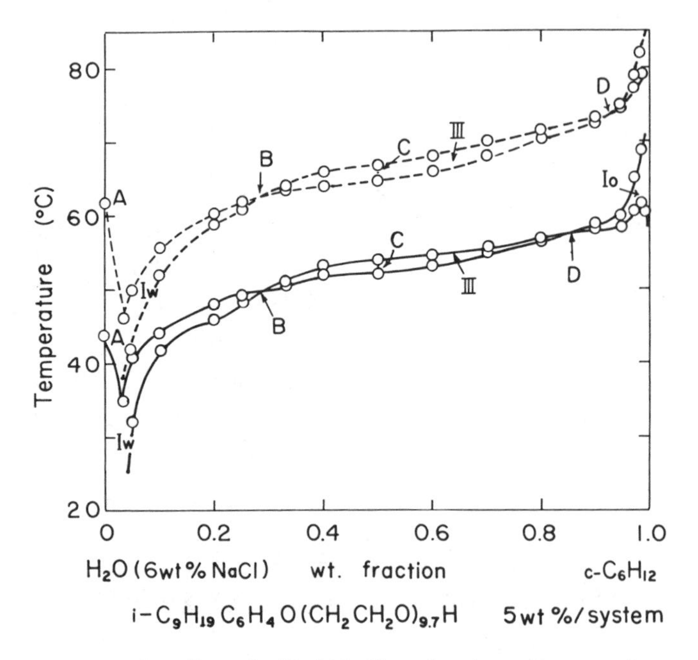

Figure 3.18 The effect of added NaCl on the phase diagram of a water/cyclohexane system containing 5 wt% of *i*-$C_9H_{19}C_6H_4O(CH_2CH_2O)_{9.7}H$ as a function of temperature. The dotted curve represents the diagram without salt and the solid curve represents that with 6 wt% NaCl in water. [Reproduced by permission of Academic, *J. Colloid Interface Sci.*, **32**, 642 (1970).]

emulsion of a 1 : 1 volume ratio, and the cloud (haze) point in nonaqueous surfactant solution saturated with water, respectively. These characteristic temperatures shift to a similar extent due to the salt added to the water. Instead of determining the effect of various amounts of added salts, acid, and alkali on the phase diagram, the effect of these additives on the PIT seems sufficient to estimate the shift of points B and D (optimum temperature for microemulsion) or phase diagram.

The effect of added salts, acid, and alkali on the PIT of emulsions containing 3 wt% of polyoxyethylene(9.7) nonylphenylether in water is shown in Figs. 3.19 and 3.20.

The relation between the PIT in 1:1 emulsions of water/cyclohexane and the HLB number of nonionic surfactants in the absence of electrolytes is plotted in Fig. 3.21. Figures 3.16 and 3.21 are useful in order to estimate the change of the HLB number and the optimum hydrophilic chain length due to the change of PIT at various temperatures. It is concluded that the presence of 6 wt% of NaCl in water corresponds to a decrease in the PIT of about 14°C or of a decrease in the HLB number of about 1.0. An emulsifier, whose PIT is higher, is required when salts are present in water to compensate for the PIT depression, because the opti-

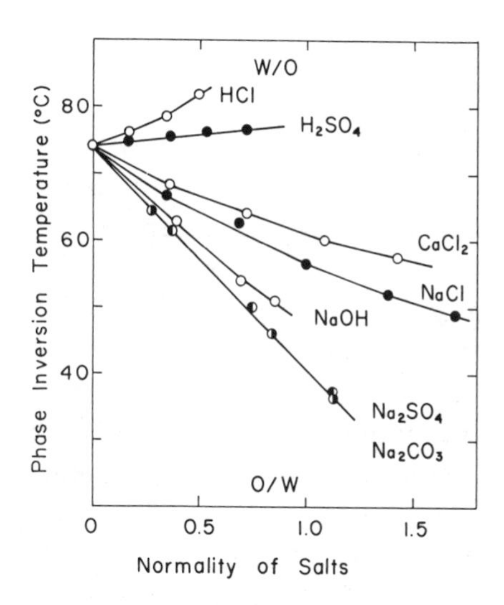

Figure 3.19 The effect of added salts, acid, and alkali on the PIT of a cyclohexane/water emulsion (1:1 volume ratio) containing 3 wt% of i-$C_9H_{19}C_6H_4O(CH_2CH_2O)_{9.7}H$ in water. (25) [Reproduced by permission of Academic, *J. Colloid Interface Sci.*, **32**, 642 (1970).]

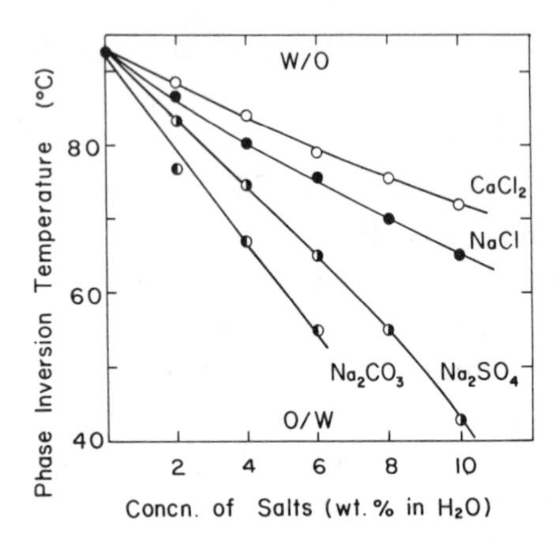

Figure 3.20 The effect of added salts on the PIT of a heptane/water emulsion (1:1 volume ratio) containing 3 wt% of $i\text{-}C_9H_{19}C_6H_4O\text{-}(CH_2CH_2O)_{9.7}H$ in water. (25) [Reproduced by permission of Academic, *J. Colloid Interface Sci.*, 32, 642 (1970).]

mum PIT of the emulsifier is fixed if the temperature is fixed (26). In other words, a so-called HLB number correction due to an added salt in water is required, in addition to the HLB number correction for the oil, in order to select an optimum emulsifier for a given system. It was difficult to determine the so-called HLB number corrections required due to the added salts in the emulsion from the usual stability versus HLB number test, because (a) the effect of added salt on the hydrophile–lipophile balance of the surfactant is not very large, and (b) the stability of the emulsion does not change a great account with the change of the hydrophilic chain length of emulsifiers (26). On the other hand, the PIT in emulsions accurately reflects the change of the HLB of the emulsifiers due to additives and affords clear information

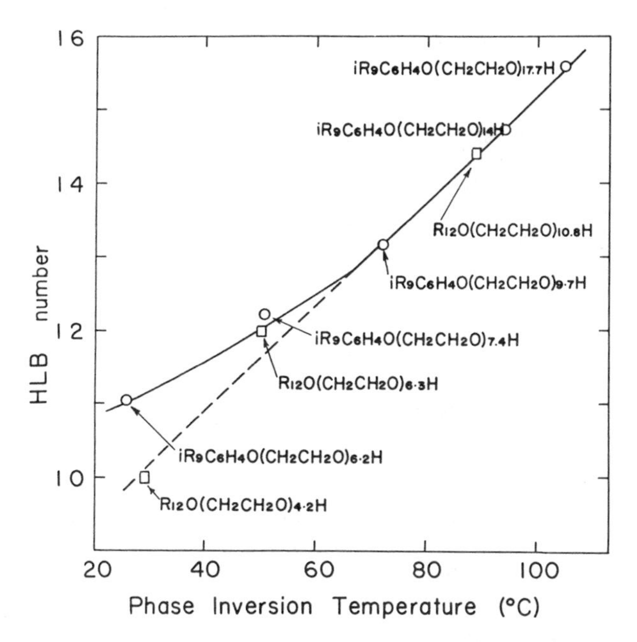

Figure 3.21 The relation between the PIT in a 1 : 1 cyclohexane/water emulsion and the HLB number of nonionic surfactants (emulsifier: 5 wt%). (25) [Reproduced by permission of Academic, *J. Colloid Interface Sci.*, **32**, 642 (1970).]

concerning the influence of the amount and kinds of additives (27–30).

If the temperature of an aqueous salt solution is higher than the cloud point in the presence of oil and lower than the PIT in an emulsion of the system, little of the nonionic emulsifier dissolves in the water phase, and the nonionic surfactant phase effectively adheres to the oil phase and disperses it into water forming an O/W-type emulsion (1). Thus, the present data are useful in selecting optimum emulsifiers for dispersing an oil in sea water or in hard water.

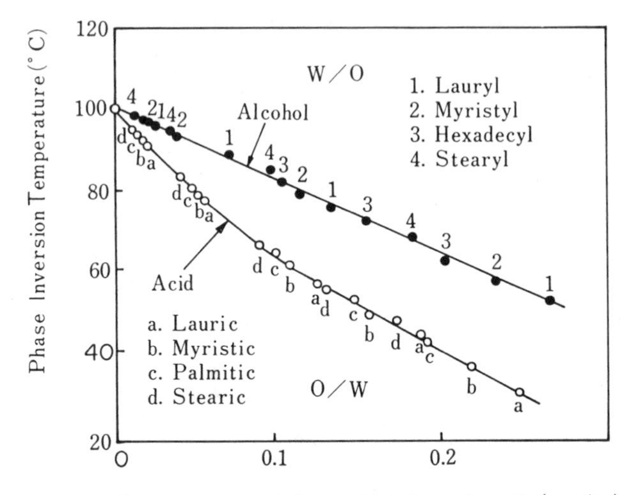

Figure 3.22 The effect of added fatty acids and alcohols on the PIT in an emulsion composed of 1 : 1 weight ratio liquid paraffin and water emulsified with 3 wt% $C_{12}H_{25}O(CH_2CH_2O)_{6.3}H$.

3.8 THE EFFECT OF ADDITIVES IN OIL ON THE PIT

The effect of added fatty acids and alcohols on the PIT in an emulsion composed of a 1 : 1 weight ratio of liquid paraffin and water emulsified with 3 wt% per system of $C_{12}H_{25}O(CH_2CH_2O)_{6.3}H$, is shown in Fig. 3.22.

The effect of fatty acids and alcohols on the PIT plotted in moles per kilogram of these additives were similar regardless of the chain length (C_{12}–C_{18}) of acids and alcohols. Thus, the effect of additives in the oil phase on the HLB of the surfactant at the interface is accurately determined by PIT measurements.

REFERENCES

1. K. Shinoda and H. Arai, *J. Phys. Chem.*, **68**, 3485 (1964).
2. T. Mitsui, Y. Machida, and F. Harusawa, *Bull. Chem. Soc. Jpn.*, **43**, 3044 (1970).
3. W. D. Bancroft, *J. Phys. Chem.*, **17**, 501 (1913); **19**, 275 (1915).
4. K. Shinoda, *J. Colloid Interface Sci.*, **24**, 8 (1967).
5. K. Shinoda and H. Arai, *J. Colloid Interface Sci.*, **25**, 429 (1967).
6. K. Sato and M. S. Yuasa, *Yukagaku*, **26**, 435 (1977) (in Japanese).
7. W. Ostwald, *Kolloid* Z., **6**, 103 (1910); **7**, 64 (1910).
8. P. Becher, *Emulsions*, 2nd ed., Reinhold, New York (1966), pp. 158–164.
9. E. H. Crook, D. B. Fordyce, and G. F. Trebbi, *J. Colloid Sci.*, **20**, 191 (1965).
10. H. L. Greenwald, E. B. Kice, M. Kenly, and J. Kelly, *Anal. Chem.*, **33**, 465 (1961).
11. F. Harusawa, T. Saito, H. Nakajima, and S. Fukushima, *J. Colloid Interface Sci.*, **74**, 435 (1980).
12. F. Harusawa and M. Tanaka, *J. Phys. Chem.*, **85**, 882 (1981).
13. H. Arai and K. Shinoda, *J. Colloid Interface Sci.*, **25**, 396 (1967).
14. W. C. Griffin, *J. Soc. Cosmet. Chem.*, **1**, 311 (1949).
15. O. Harva, P. Kivalo, and A. Keltakallio, *Suom. Kemistilehti B*, **32**, 52 (1959).
16. N. Ohba, *Bull. Chem. Soc. Jpn.*, **35**, 1016 (1962).
17. V. R. Huebner, *Anal. Chem.*, **34**, 488 (1962).
18. W. G. Gorman and G. D. Hall, *J. Pharm. Sci.*, **52**, 442 (1963).

19. P. Becher and R. L. Birkmeier, *J. Am. Oil Chem. Soc.*, **41**, 169 (1964).
20. W. C. Griffin, *J. Soc. Cosmet. Chem.*, **5**, 249 (1954).
21. J. H. Schulman and E. G. Cockbain, *Trans. Faraday Soc.*, **36**, 651 (1940).
22. K. Shinoda, H. Saito, and H. Arai, *J. Colloid Interface Sci.*, **35**, 624 (1971).
23. E. H. Crook and D. B. Fordyce, *J. Am. Oil Chem. Soc.*, **41**, 231 (1964).
24. K. Shinoda, T. Yoneyama, and H. Tsutsumi, *J. Dispersion Sci. Technol.*, **1**, 1 (1980).
25. K. Shinoda and H. Takeda, *J. Colloid Interface Sci.*, **32**, 642 (1970).
26. K. Shinoda and H. Saito, *J. Colloid Interface Sci.*, **30**, 258 (1969).
27. L. Marszall, *Fette Seifen Anstrichm.*, **79**, 41 (1977).
28. L. Marszall, *Fette Seifen Anstrichm.*, **80**, 289 (1978).
29. L. Marszall, *J. Colloid Interface Sci.*, **59**, 376 (1977); **60**, 570 (1977).
30. L. Marszall and J. W. van Valkenburg, in *Advances in Pesticide Science* (H. Beissbuhler, Ed.), Vol. 3, Pergamon, Oxford (1979), p. 789.

CHAPTER 4

Stability of Emulsion

4.1 INITIAL DROPLET DIAMETER AND STABILITY OF O/W-TYPE EMULSIONS AS FUNCTIONS OF TEMPERATURE AND OF PIT (HLB TEMPERATURE) OF EMULSIFIERS

In Chapter 1, the phase diagram and the dispersion types of water/cyclohexane systems containing 3–7 wt% of polyoxyethylene(9.7)nonylphenylether were treated as a function of temperature (1). In this chapter, we discuss: (a) the mean droplet diameter and the stability of emulsions as functions of temperature and the hydrophile–lipophile balance (HLB) of the emulsifier, and (b) the correlations among the optimum temperature for stable emulsification, the optimum hydrophilic chain length, and the phase inversion temperature, PIT, of emulsifiers.

The effect of emulsification temperature on the mean volume diameter of O/W-type emulsions of the ternary system composed of 3 wt% of polyoxyethylene(9.7)nonylphenylether, 48.5 wt% of water, and 48.5 wt% of cyclohexane are shown in Fig. 4.1a (upper figure). The volume fractions of oil, cream, and water phases of the same system 5 h after agitation are shown in Fig. 4.1b (lower figure). Cream phase means an O/W-type emulsion below the PIT and a W/O-type emulsion above the PIT. [The PIT of this system was 72°C (2).]

As can be seen in Fig. 4.1a, the mean volume diameter right after the emulsification is smallest close to the PIT and monotonically increased with the temperature decrease, reflecting the change of the oil/water interfacial tension and the fraction of surfactant phase being present in the system as a function of temperature (3–6). In the course of time, the droplet diameter increases rapidly close to the PIT, because in that range, coalescence is facilitated due to the ultralow

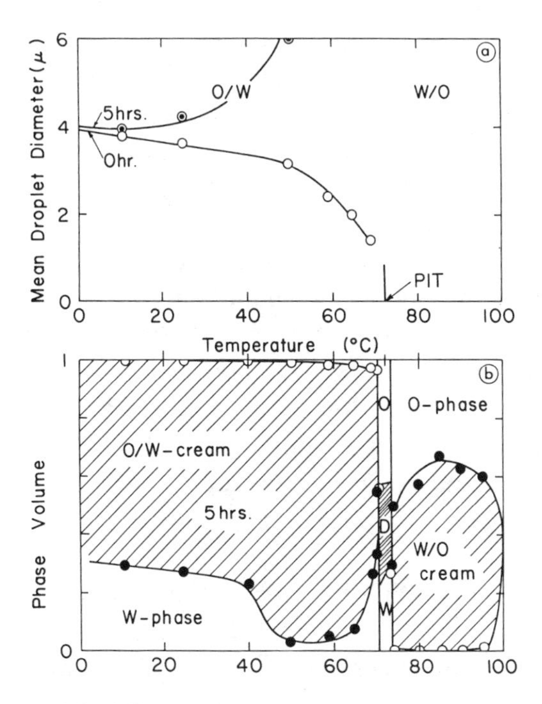

Figure 4.1 (a) The effect of temperature on the mean volume diameter of emulsions containing 3 wt% polyoxyethylene(9.7)nonylphenylether, 48.5 wt% cyclohexane, and 48.5 wt% water immediately after (0 h) and 5 h after emulsification. (b) The effect of temperature on the volume fractions of oil, cream, and water phases in the same system. After emulsification with a single emulsifier the system was maintained at the temperature indicated for 5 h. [Reproduced by permission of Academic, New York, *J. Colloid Interface Sci.*, **30**, 258 (1969).]

interfacial tension (4,5). It increases more slowly at lower temperatures, because the coalescence rate is now lower. It is evident from Fig. 4.1b that the mean volume diameter, coalescence rate, and so on, of an emulsion stabilized with a definite nonionic emulsifier varies remarkably with temperature. As for the O/W-type emulsion, the drainage rate was slow at about 20–40°C below the PIT, but for the W/O-type emulsion it was slow at about 20°C above the PIT. As the PIT varies with the types of oils and the hydrophilic chain lengths of nonionic emulsifiers, it is readily understood that the optimum PIT, that is, the optimum hydrophilic chain length of the emulsifier, is required to get a better emulsion for respective oils and temperature (2).

In a similar manner, an alteration of the hydrophilic chain length of the emulsifier should have an affect similar to the temperature change of the system since the interaction between the hydrophilic moiety and water is modified with the temperature variation. The effect of the hydrophilic chain length of a series of nonionic agents on the mean volume diameter and the stability of emulsions of an O/W-type emulsion support this reasoning as shown in Fig. 4.2a and 4.2b (2).

The higher the PIT of the emulsifier is than the experimental temperature, the larger the emulsion droplets will be. This finding is consistent with the results obtained in Fig. 4.1, that is, the lower the emulsification temperature, the larger the droplet diameter of a system stabilized with a definite emulsifier. It is evident from Fig. 4.2 that emulsifiers, the PITs of which are about 30–70°C higher than storage temperature, yield the most stable emulsions. Hence, the screening for a suitable emulsifier for given ingredients at a given temperature by the use of the PIT data is a straightforward and simple process.

Emulsions are most unstable against coalescence at the PIT, due to the ultralow interfacial tension (4,5). PIT can be accurately determined, whereas the maximum stability of the emulsion rather insensitively changes with the hydrophilic chain length of the emulsifiers (2) such that the optimum HLB number for most stable emulsifiers cannot be determined precisely.

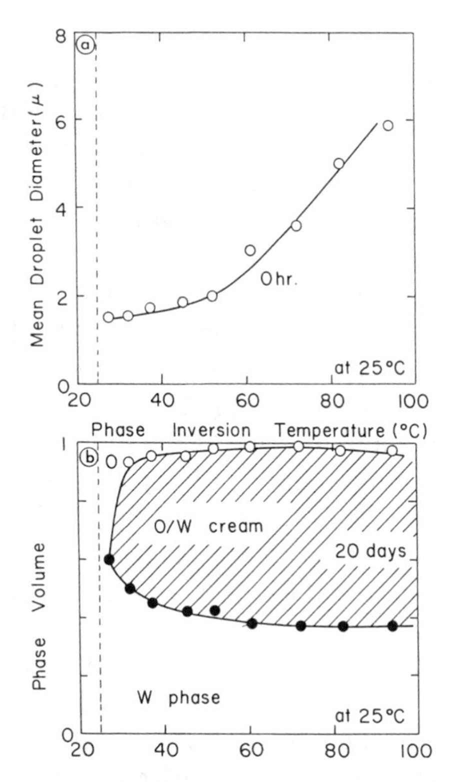

Figure 4.2 (a) The effect of PIT (hydrophilic chain length) of emulsifiers on the mean volume diameter of emulsions containing 3 wt% polyoxyethylene(6.3–14.0) nonylphenylethers, 48.5 wt% cyclohexane, and 48.5 wt% water. (b) The effect of PIT of an emulsifier on the volume fraction of oil, cream, and water phases 20 days after emulsification. (Emulsified and stored at 25°C using a series of polyoxyethylene nonylphenylethers the PITs of which vary from 27–94°C.) [Reproduced by permission of Academic, New York, *J. Colloid Interface Sci.*, **30**, 258 (1969).]

4.1.1 Emulsification by PIT Method

It is readily anticipated from the experiments shown in Figs. 4.1 and 4.2 that a better emulsion can be obtained if a system is initially emulsified close to the PIT (about 2–4°C below the PIT appears optimal) in order to primarily obtain a fine dispersion at which interfacial tension is small (4) and then rapidly cooled down to the storage temperature, at which the coalescence rate is slow. The cooling process adds small droplets from the incipient phase separation of the surfactant phase (6). We designate this process as "emulsification by the PIT method" (2). The effect of emulsification temperature on the mean volume diameter and the stability of emulsions as a function of emulsification temperature have been observed and are shown in Fig. 4.3 (2).

It is clear that emulsification at the PIT (slightly below) affords smaller droplets of the O/W-type emulsion. On the contrary, emulsification at a temperature higher than the PIT means emulsification to a W/O-type emulsion at first and then inversion to an O/W-type emulsion by cooling.

The latter process, "emulsification by the inversion method," does not result in such small droplets as "emulsification by the PIT method,"' since the small droplets from the surfactant phase are not obtained (6). In the "emulsification by the inversion method," an emulsifier is dissolved in the oil phase. The emulsion may then be formed by adding water directly to the mixture. In this case, a W/O-type emulsion is formed at first and then inverted to an O/W type by the further addition of water. On the other hand, in "emulsification by the PIT method," emulsifier, oil, and a smaller amount of (a part of) water are mixed at the same time at slightly below the PIT. An O/W-type emulsion may be formed at this temperature, and the emulsion is subsequently cooled down to the storage temperature by adding cool water. The surfactant phase, which is found at the HLB temperature, will separate into two phases (surfactant and oil phase and the

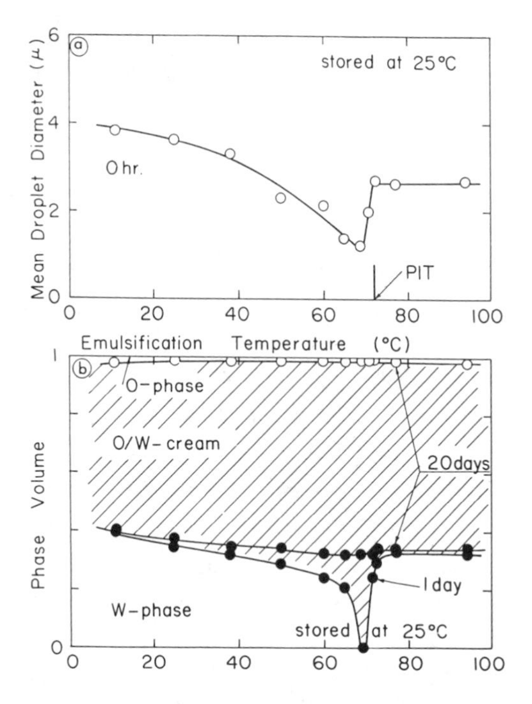

Figure 4.3 (a) The effect of the emulsification temperature on the mean volume diameter of emulsions containing 3 wt% polyoxyethylene(9.7) nonylphenylether, 48.5 wt% cyclohexane, and 48.5 wt% water. (b) The effect of the emulsification temperature on the volume fractions of oil, cream, and water phases of the same emulsion 20 days after agitation. After emulsification with a single emulsifier the system was stored at 25°C. [Reproduced by permission of Academic, New York, *J. Colloid Interface Sci.*, **30**, 258 (1969).]

former continuously change to aqueous micellar solution) when the temperature is reduced. This phase separation will spontaneously give extremely small droplets during the cooling process. These extremely small droplets skew the average droplet size to smaller values (6).

4.1.2 Comparison of Emulsions Prepared by Simple Shaking and Those by the PIT Method

The mean volume diameters of emulsions emulsified at the PIT and cooled down to various storage temperatures, as well as the phase volumes of oil, cream, and water phases of the same system, are shown in Fig. 4.4. The mean volume diameter of emulsions made at several temperatures without cooling is also shown for comparison by the dotted line. It is evident from Fig. 4.4 that: (a) the smaller droplets are obtained by emulsification at the PIT regardless of the storage temperature and thus (b) emulsions stored at lower temperatures, at which the coalescence rate is slow, are more stable.

Using a series of polyoxyethylene nonylphenylethers, whose PITs change over a wide temperature range, the emulsification by the PIT method for the cyclohexane/water system was studied at respective PITs. The mean volume diameter and phase volumes of the respective phases as a function of the PITs of emulsifiers are plotted in Fig. 4.5 (2).

The initial droplet was equally small regardless of the PITs, whereas the emulsions shaken and stored at 25°C show a gradual increase of droplet size with the PIT rise, as shown in Fig. 4.2. It is concluded from Fig. 4.5 that emulsifiers, the PITs of which are about 30–65°C higher than the storage temperature, provide optimally stable emulsions.

The PIT change from 27–94°C for a series of emulsifiers in this experiment corresponds to the change of HLB number for them from 11.1–14.7. The difficulties with an accurate determination of the optimum HLB number from the stability versus HLB-number relation are obvious from Fig. 4.5b; the maximum stability is insensitive to the change in HLB number. On the other hand, the instability of the emulsion is sensitive to the PIT, so that the selection of a suitable emulsi-

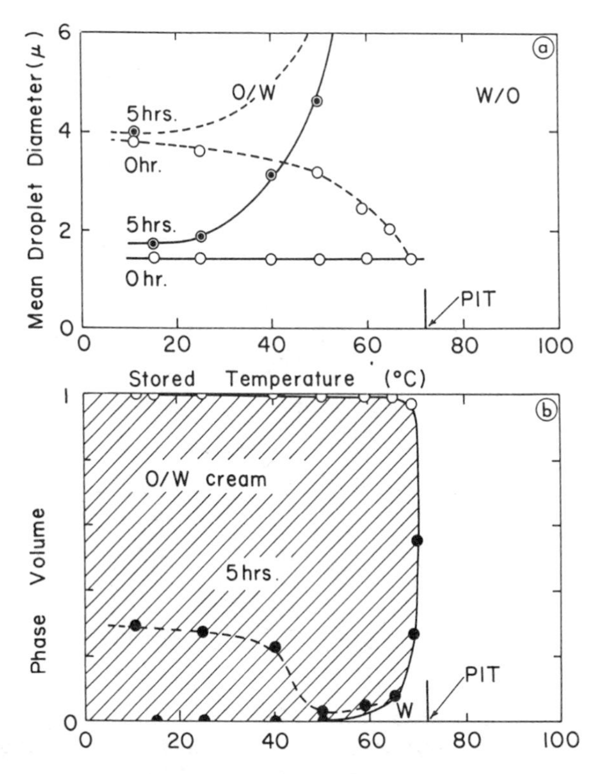

Figure 4.4 (a) The mean volume diameter of O/W-type emulsions [3 wt% polyoxyethylene(9.7) nonylphenylether, 48.5 wt% cyclohexane, and 48.5 wt% water] emulsified at the PIT and then cooled to and held at the temperature indicated. Holding times were 0 and 5 h. Dotted curve illustrates the mean droplet diameter of the same system emulsified and stored at one temperature. (b) The effect of storage temperature on the volume fraction of the oil, cream, and water phases of the system prepared as in Fig. 4.4a and stored 5 h. [Reproduced by permission of Academic, New York, *J. Colloid Interface Sci.*, **30**, 258 (1969).]

fier by the PIT data is accurate. The optimum temperature difference between the storage temperature and the PIT is the only variable; as long as it is in the range 30–65°C the emulsifier will perform well.

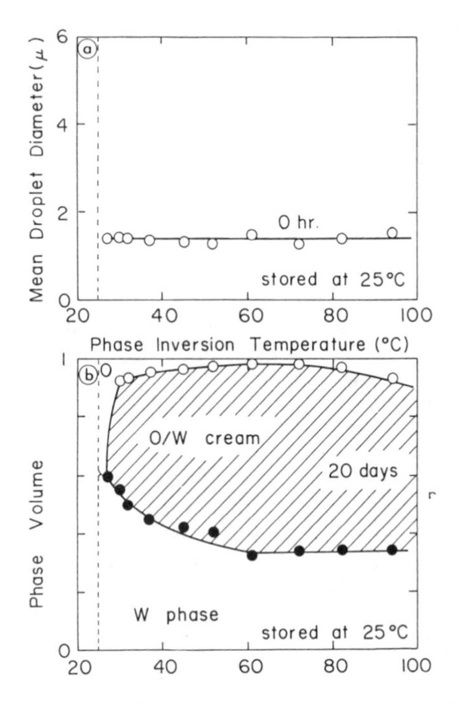

Figure 4.5 (a) The effect of the PIT, that is, the hydrophilic chain length, of the emulsifiers on the mean volume diameter of emulsions containing 3 wt% of polyoxyethylene nonylphenylethers, 48.5 wt% cyclohexane, and 48.5 wt% water. (b) The volume fractions of the oil, cream, and water phases of the above systems 20 days after agitation (emulsified at the PIT and stored at 25°C). [Reproduced by permission of Academic, New York, *J. Colloid Interface Sci.*, **30**, 258 (1969).]

The comparison between the phase contrast microscopic photographs of emulsions prepared by the PIT method and by a simple shaking is shown in Fig. 4.6. The system contains 3 wt% polyoxyethylene(8.6)nonylphenylether, 77.6 wt% water, and 19.4 wt% cyclohexane.

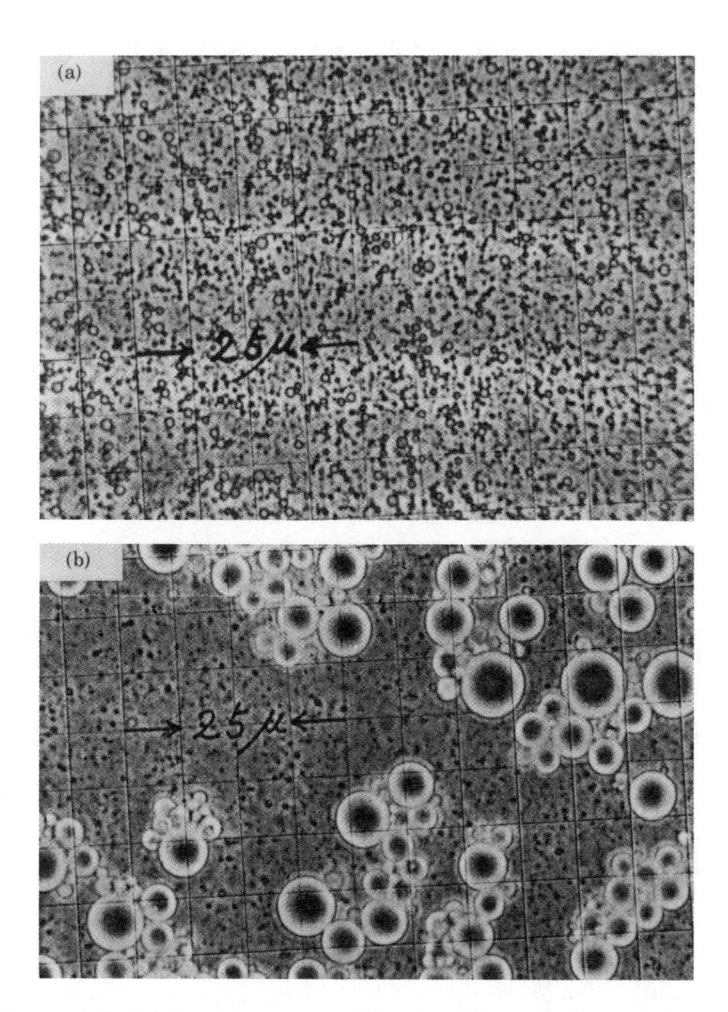

Figure 4.6 (a) Emulsion prepared by the PIT method, emulsified at 49°C, and cooled down to 25°C. The PIT of this system was 52°C. (b) Emulsion prepared by simple shaking at 25°C. [Reproduced by permission of Academic, New York, *J. Colloid Interface Sci.*, **30**, 258 (1969).]

4.2 EMULSIFIER SELECTION AND THE STABILITY OF W/O-TYPE EMULSIONS AS FUNCTIONS OF TEMPERATURE AND HYDROPHILIC CHAIN LENGTH OF EMULSIFIER

Water/oil-type emulsions have been found most stable against coalescence for the following combinations of storage temperatures and PIT values. The storage temperatures were 87 and 53°C for emulsifiers with PITs, 72 and 25°C in order to obtain optimally stable W/O-type emulsions (2,3). Hence, for a storage at room temperature, the PIT of a suitable emulsifier to obtain a stable W/O-type emulsion may be lower than 0°C and cannot be determined. Hence, the election of a W/O-type emulsifier must be pursued using a different method than that used for the O/W-type emulsifier. The attention is focused on the influence of the hydrophilic chain length of the emulsifier.

The effect of temperature on the stability of cyclohexane—water emulsions stabilized with 3 wt% per system of $C_9H_{19}C_6H_4O—(CH_2CH_2O)_{7.4}H$ is shown in Fig. 4.7 (7).

The temperature range of a relatively stable W/O-type emulsion, in which the coalesced water phase was small one day after agitation, was observed 10–30°C above the PIT (51°C) in this system in accordance with earlier results (2,3). The emulsion was unstable close to the PIT, and the system separated into three phases, that is, water, surfactant, and oil phases. The aqueous phase observed at lower temperatures was converted into the surfactant phase in the three-phase region and then converted into the oil phase at temperatures above the three-phase region as shown in Fig. 4.7. This phenomenon is illustrated schematically in Fig. 4.8 (7).

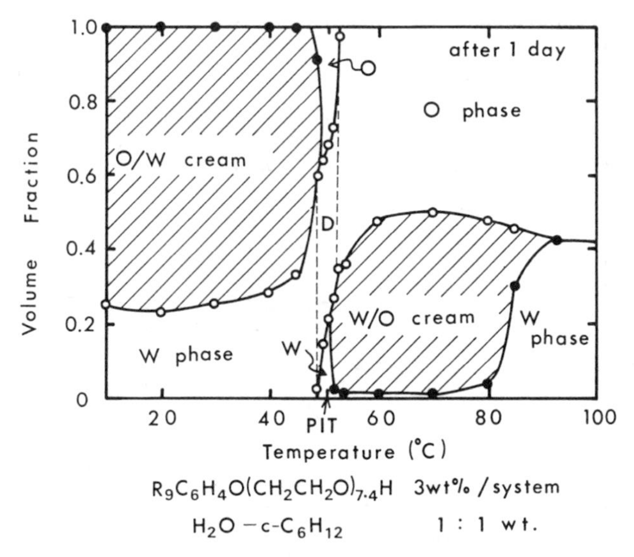

Figure 4.7 The effect of temperature on the types of emulsions, and the volume fractions of oil, cream, and water phases 1 day after agitation. (○) Drained phase-cream boundary; (●) coalesced phase-cream boundary. [Reproduced by permission of Academic, New York, *J. Colloid Interface Sci.*, **64**, 68 (1978).]

A more complete description of the complicated phase changes was given in Chapter 1. These phase changes as functions of temperature are parallel to phase changes induced by a change in the polar chain length at constant temperature (8). This means that more hydrophilic emulsifiers with larger polar chains correspond to temperatures below the HLB temperature and more lipophilic ones to the range above. At the HLB temperature the hydrophilic–lipophilic properties balance.

Hence, the temperature effect can be replaced by a polar chain length effect. The effect of the hydrophilic chain length of emulsifiers on the types and stability of cyclohexane/

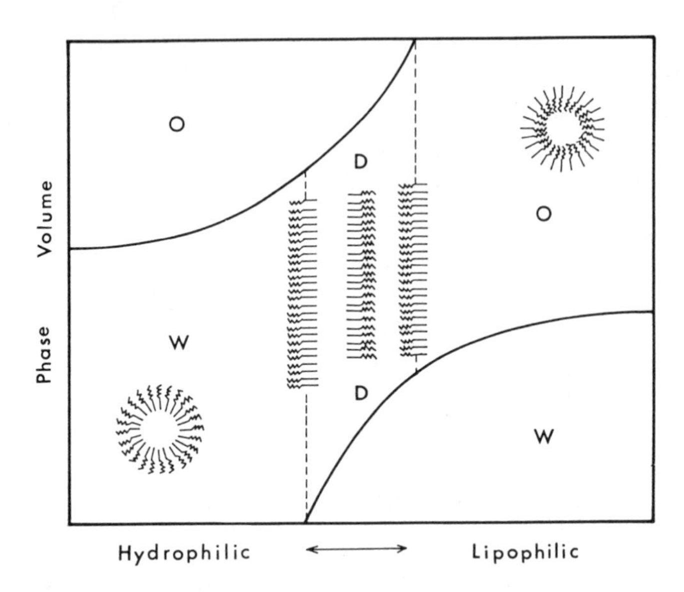

Figure 4.8 Schematic diagram of water, surfactant, and oil phases after the complete phase separation of the system shown in Fig. 4.7. Surfactant is lipophilic at higher temperature. [Reproduced by permission of Academic, New York, *J. Colloid Interface Sci.*, **64**, 68 (1978).]

water emulsions stabilized with $C_9H_{19}C_6H_4O(CH_2CH_2O)_nH$ at 25°C is shown in Fig. 4.9 (7).

The PIT in Fig. 4.9 is shown to be 25°C for an emulsion emulsified with $C_9H_{19}C_6H_4O(CH_2CH_2O)_{6.2}H$. More lipophilic emulsifiers formed W/O-type emulsions, whereas more hydrophilic emulsifiers formed O/W-type emulsions. From these data the stability contours of emulsions as functions of temperature and the oxyethylene chain length of nonionic emulsifiers on the stability of 1:1 oil/water emulsions containing 3–5 wt% of emulsifiers were constructed for liquid paraffin/water and cyclohexane/water systems and shown in Figs. 4.10 and 4.11 (7).

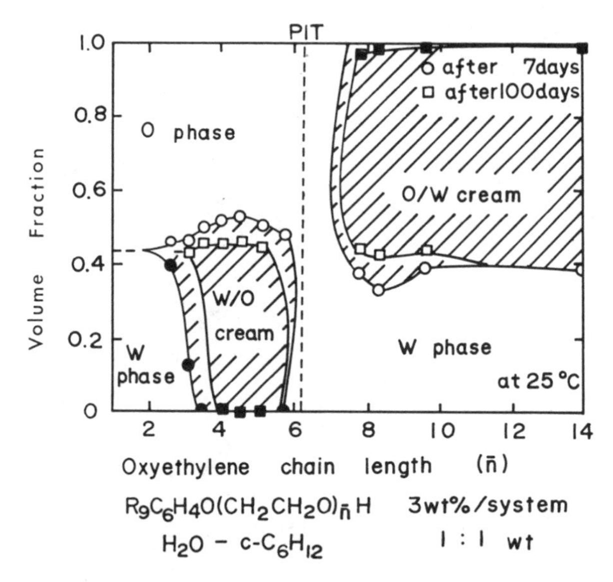

Figure 4.9 The effect of the hydrophilic chain length of emulsifiers on the types and stability of cyclohexane/water emulsions stabilized with 3 wt% of $C_9H_{19}C_6H_4O(CH_2CH_2O)_nH$ at 25°C. (○, □) drained phase-cream boundary; (●, ■) coalesced phase-cream boundary. [Reproduced by permission of Academic, New York, *J. Colloid Interface Sci.*, **64**, 68 (1978).]

The slope of the PIT versus the oxyethylene chain length of emulsifiers is very steep at 0–25°C for both oils. It is evident from the stability contours that the hypothetical PIT of emulsifiers suitable for stable W/O-type emulsions at 25°C is far below 0°C.

The W/O-type emulsions were most stable for both cases of the oxyethylene chain length of emulsifier about 0.70–0.88 times that of the emulsifier whose PIT (HLB temperature) is equal to the storage temperature. For the emulsion at 25°C, liquid paraffin/water emulsions would require

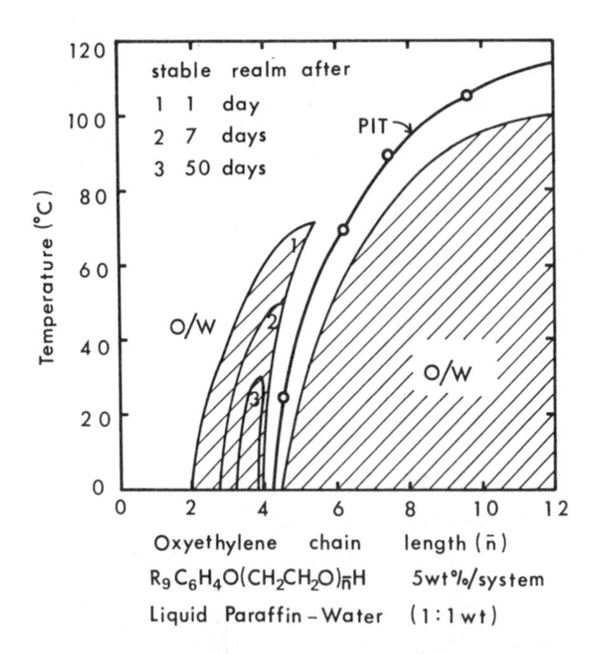

Figure 4.10 The stability contour of liquid paraffin/water emulsions stabilized with 5 wt% of $C_9H_{19}C_6H_4O(CH_2CH_2O)_nH$ as functions of temperature and oxyethylene chain length of emulsifiers in which the coalesced phase was not observed. [Reproduced by permission of Academic, New York, *J. Colloid Interface Sci.*, **64**, 68 (1978).]

$R_9C_6H_4O(CH_2CH_2O)_{3.2-4.4}H$ and cyclohexane/water emulsions would require $R_9C_6H_4O(CH_2CH_2O)_{4.3-5.6}H$.

It is evident from Figs. 4.10 and 4.11 that: (a) the PIT is an important characteristic temperature at which the emulsion is unstable and (b) the HLB number of an emulsifier whose HLB temperature (PIT) is equal to the storage temperature (25°C), is a balanced HLB number for respective oils. The HLB numbers required to form stable W/O- or O/W-type emulsions are smaller or larger than the balanced

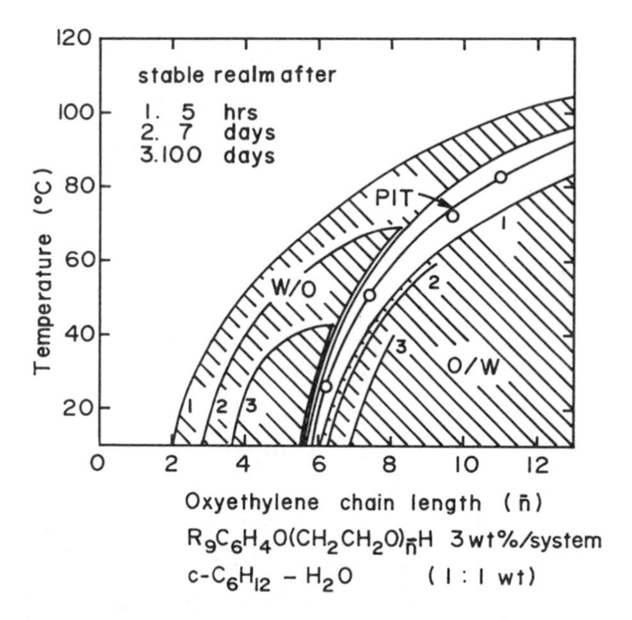

Figure 4.11 The stability contour in which the coalesced phase was not observed, of cyclohexane/water emulsions stabilized with 3 wt% of $C_9H_{19}C_6H_4O(CH_2CH_2O)_nH$ as functions of temperature and oxyethylene chain length of emulsifiers. [Reproduced by permission of Academic, New York, *J. Colloid Interface Sci.*, 64, 68 (1978).]

HLB number for oil, and a comparison with Atlas required HLB numbers is a useful exercise.

Based on a vast amount of experience, Griffin (9,10) has found that all oils, waxes, and other materials incorporated into emulsions have an individual required HLB number. This means that an emulsifier, or blend of emulsifiers, having a required HLB number ($\pm$ 1) will make a more stable emulsion than emulsifiers of another HLB number. From data on the required HLB number to emulsify various oil phases, cyclohexane requires 15 for the O/W type (11), paraffinic mineral oil requires 10–12 for the O/W type, and 4 for the W/O-type emulsion (10,12).

The HLB numbers of emulsifiers that yielded most stable W/O-type emulsions in the present study were calculated using Eq. (1) (9,12).

$$\text{HLB number} = \frac{E}{5} \qquad (1)$$

where E is the weight percentage of the oxyethylene content in emulsifiers where only ethylene oxide is used as the hydrophilic portion.

It is now possible to make a comparison between the HLB number and HLB temperature (PIT) system based on the fact that the HLB of emulsifier balances at the PIT. Calculating the HLB number of the emulsifier whose PIT is equal to 25°C (Figs. 4.10 and 4.11) from the molecular formula of the emulsifier and taking into account that an O/W-type emulsion is most stable against coalescence when the PIT of the emulsifier is 20–55°C higher than the storage temperature, 25°C (13), an evaluation can be made of the range of oxyethylene chain lengths and HLB numbers of emulsifiers that yield the most stable O/W-type emulsions with the aid of Figs. 4.10 and 4.11. The required HLB numbers for the PIT systems are calculated using the fact that the W/O-type emulsions were most stable when the oxyethylene chain length of emulsifier is about 0.70–0.88 times that of the emulsifier with a PIT that is equal to the storage temperature, 25°C. These HLB numbers are summarized and compared with the HLB number recommended by Atlas in Table 4.1 (11).

The agreement is fairly good for O/W-type emulsions, but Atlas required HLB numbers for W/O-type emulsions are always lower than those evaluated from the HLB number of the balanced surfactant.

TABLE 4.1 Comparison between the Atlas Required HLB Numbers and Those Obtained from the PIT (Shinoda's HLB Numbers)

	Required HLB Number W/O		HLB Number or Balanced Emulsifier	Required HLB Number O/W	
Kinds of Oils	(Atlas)	(PIT)	(PIT)	(Atlas)	(PIT)
Liquid paraffin (paraffinic) hexadecane	4 ± 1	7.8–9.0	9.5	10 ± 1	10.00–11.7
$(CH_3)_2C_6H_4$			10.1		13–14
Cyclohexane		9.2–10.6	11.1	15 ± 1	11.7–13.5
Toluene			12.1		
CCl_4			13.0		

This discrepancy may be ascribed to several factors such as the need of more lipophilic emulsifier when a *mixed* oxyethylene chain length distribution is encountered in a mixture of commercial emulsifiers. This is especially so where different types of emulsifiers (ionic, nonionic, and zwitterionic) are combined.

4.3 THE EFFECT OF THE SIZE OF EMULSIFIER AND THE DISTRIBUTION OF THE OXYETHYLENE CHAIN LENGTH OF NONIONIC EMULSIFIERS ON THE STABILITY OF EMULSIONS

The relation between emulsion stability and the distribution of the hydrophilic chain lengths is an important topic of interest to everyone concerned with emulsion stability (14). Mayhew and Hyatt found that the undistilled material gave more stable emulsions than the molecularly distilled polyoxyethylene nonylphenylether (15). Crook and Fordyce found also that emulsions made with the normal distribution polyoxyethylene octylphenylether were more stable than those made with the corresponding homogeneous compounds (16). Hence, it is very helpful if we know the optimum distribution of the hydrophilic chain length to yield the most stable emulsion. The stability of emulsions stabilized with nonionic surfactants changes remarkably at the PIT. The PIT also changes sensitively with the hydrophilic chain length of the emulsifier (17,18) and the O/W-type emulsion is most stable when the PIT of the system is about 30–60°C higher than the storage temperature. Hence, nonionic emulsifiers having the same PIT have to be used to know: (a) the effect of the size of the hydrophilic and lipophilic groups and

(b) the effect of the distribution of the polyoxyethylene chains, and so on, on the stability of emulsions stabilized with nonionic emulsifiers. The HLB number is generally too insensitive for this purpose (14).

4.3.1 The Effect of the Size of the Hydrophilic and Lipophilic Moieties on the Stability of Emulsions

In order to determine the effect of the size of the emulsifier, it is necessary to select emulsifiers with the same PIT, that is, HLB number. However, it is difficult to select a series of nonionic emulsifiers, whose HLBs are very close to each other. Because the stability of emulsions does not change sensitively around the optimum HLB number (2), the determination of an HLB number or the selection of an optimum emulsifier from a respective series of emulsifiers is less accurate. On the other hand, the PIT is an extremely sensitive property (2) and polyoxyethylene alkyl(C_6–C_{16})phenylethers with PITs close to 60°C were chosen for a study of the importance of size of the lipophilic and hydrophilic parts when they are balanced. Systems composed of 48.5 wt% of water, 48.5 wt% of cyclohexane, and 3 wt% of nonionic emulsifiers were well shaken about 1–3°C below the PIT and then rapidly cooled down to 25°C and kept over 60 days. The emulsion droplets prepared by the PIT method are small and relatively uniform in size (2), so that the stability was compared for the emulsions prepared by this method whenever possible. The results are summarized in Table 4.2 (14).

It is evident that the stability for coalescence of O/W-type emulsions remarkably increases with the size of the hydrophile–lipophile group.

Similar experiments carried out for W/O-type emulsions

TABLE 4.2 The Effect of the Sizes of Hydrophile and/or Lipophile Groups of Polyoxyethylene Alkylphenylethers on the Stability of O/W-Type Emulsions[a]

Emulsifier	HLB Number	PIT (°C)	Mean Droplet Diameter[b] (μ)		Volume Percent after 60 days		Relative Stability for Coalescence[c]
			0 day	60 days	Drained Phase	Coalesced Phase	
$R_6C_6H_4O(CH_2CH_2O)_{7.5}H$	13.0	65	1.4	15	50	8	0.5
$R_8C_6H_4O(CH_2CH_2O)_{8.4}H$	12.8	60.6	1.5	3.6	42	5	0.8
$R_9C_6H_4O(CH_2CH_2O)_{8.6}H$	12.6	60.5	1.6	3.5	40	3	1
$R_{12}C_6H_4O(CH_2CH_2O)_{9.7}H$	12.4	61	1.4	2.0	25	2	10
$R_{16}C_6H_4O(CH_2CH_2O)_{12.4}H$	12.6	49.5	1.6	1.7	16	1	>100

[a]The emulsions are composed of 48.5 wt% of water, 48.5 wt% of cyclohexane, and 3 wt% of nonionic emulsifiers. The systems were shaken at about 1–3°C below the PIT, cooled, and stored at 25°C over 60 days.

[b]Mean volume diameter, $D = \sqrt[3]{\Sigma_i n_i D_i^3 / \Sigma_i n_i}$, of emulsion droplets in the cream phase.

[c]Relative stability means approximate rate of coalescence of emulsion droplets relative to polyoxyethylene nonylphenylether.

TABLE 4.3 The Effect of the Sizes of Hydrophile and/or Lipophile Groups on the Stability of W/O-Type Emulsions[a]

Emulsifier	HLB Number	PIT (°C)	Volume Percent after 30 days	
			Drained Phase	Coalesced Phase
$R_6C_6H_4O(CH_2CH_2O)_{7.5}H$[b]	13.0	52	68	32
$R_8C_6H_4O(CH_2CH_2O)_{8.5}H$	12.9	57	68	32
$R_9C_6H_4O(CH_2CH_2O)_{8.4}H$	12.4	50	65	27
$R_{12}C_6H_4O(CH_2CH_2O)_{9.4}H$	12.2	51	63	0
$R_{16}C_6H_4O(CH_2CH_2O)_{12.4}H$	12.6	48	59	0

[a]The emulsions are composed of 47.5 wt% water, 47.5 wt% of cyclohexane, and 5 wt% of nonionic emulsifiers. The systems were shaken and stored at 64°C over 30 days.

[b]In this case, 10 wt% of emulsifier was added instead of 5 wt%.

are summarized in Table 4.3. The stability of W/O-type emulsions also increases rapidly with the size of the emulsifiers (14).

4.3.2 The Effect of the Distribution of the Hydrophilic Chain Lengths of Emulsifiers on the Stability of Emulsions

It is evident from the preceding sections that: (a) the stability of an emulsion changes remarkably with temperature and with the hydrophilic chain length of the emulsifier and (b) the effect of the temperature rise and the effect of the decrease of the hydrophilic chain length are quite similar. Accordingly, it is necessary to choose nonionic emulsifiers of the same PIT (HLB) in order to determine such delicate relations as that between the emulsion stability and the distribution of the hydrophilic chain length of the emulsifier. It is also evident that raising the storage temperature in order to accelerate the stability test of emulsions stabilized with nonionic surfactant invalidates the test.

Such an interesting and valuable subject as the effect of the distribution of the hydrophilic chain length on the stability of emulsions has been investigated by many workers; however, the experiments carried out at elevated temperatures to accelerate the stability test, or those studies using emulsifiers, the PITs of which are not the same, are meaningless. These difficulties have been conquered by: (a) the PIT system for an accurate HLB (PIT) determination, (b) the PIT method for emulsification, (c) the synthesis of pure nonionics, and (d) progress in the analysis of the distribution of polyoxyethylene chain length (19).

The relation between the emulsion stability and the distribution of hydrophilic chain length has been compared for emulsions stabilized with homogeneous, molecularly distilled and commercial polyoxyethylene dodecylether. The results are summarized in Table 4.4 and the distribution of the oxyethylene chain lengths is plotted in Fig. 4.12 (14).

It has to be kept in mind that the sharply distributed materials appear more hydrophilic, as shown by higher cloud points, than broader samples of the same average polyoxyethylene chain length (14), because the small amount of short-chain material effectively depresses the cloud point (20). On the other hand, the broadly distributed materials appear more hydrophilic as shown by higher PITs than samples of the sharply distributed materials, because the shorter chain homolog preferentially dissolves into the oil phase and the average polyoxyethylene chain lengths of the adsorbed materials at the interface become more hydrophilic. Actually, samples 1, 2, and 4 in Table 4.4 show the same PIT but different oxyethylene chain lengths (HLB number), whereas the average oxyethylene chain length of sample 3 is close to those of samples 1 and 2, but showing the higher PIT. It is evident from Table 4.4 and Fig. 4.12 that the emulsion made with homogeneous surfactant was the most unstable and that with the commercial material, whose distribution of oxyethylene chain is broad, was much more stable. For example, the mean droplet diameters after 4-months storage were 70 μ for the homogeneous, 30 μ for the molecularly distilled, and 5 μ for the commercial materials. It is known that a mixture of lipophilic and hydrophilic emulsifiers yields better emulsions than simple emulsifiers (21). An expected and probable relation is to the packing at the interface; a mixture shows a lower surface tension than the pure compound (22) at the air/solution interface showing enhanced adsorptions.

Similar experiments carried out for systems stabilized with various polyoxyethylene nonylphenylethers whose PITs are close to 60°C, are summarized in Table 4.5. The distribution

TABLE 4.4 The Effect of the Distribution of the Hydrophile Chain Length of Polyoxyethylene Dodecylether on the Stability of O/W-Type Emulsions[a]

Emulsifier	HLB Number	PIT (°C)	Mean Droplet Diameter (μ) 4 months	Volume Percent after 4 months		Relative Stability for Coalescence
				Drained Phase	Coalesced Phase	
1. $R_{12}O(CH_2CH_2O)_8H$ (homogeneous)	13.1	51.5	70	40	5	1/3
2. $R_{12}O(CH_2CH_2O)_{8.2}H$ (molecularly distilled	13.2	53.5	30	38	4.5	1 (standard)
3. $R_{12}O(CH_2CH_2O)_{8.0}H$ (mixture of molecularly distilled)	13.1	64.7	15	32	3.5	3
4. $R_{12}O(CH_2CH_2O)_{5.3}H$ (commercial)	11.1	54	5	32	3.5	5

[a]The emulsions are composed of 48 wt % of water, 48 wt % of cyclohexane, and 4 wt % of nonionic emulsifiers. The systems were shaken and stored at room temperature (10–35°C) over 4 months.

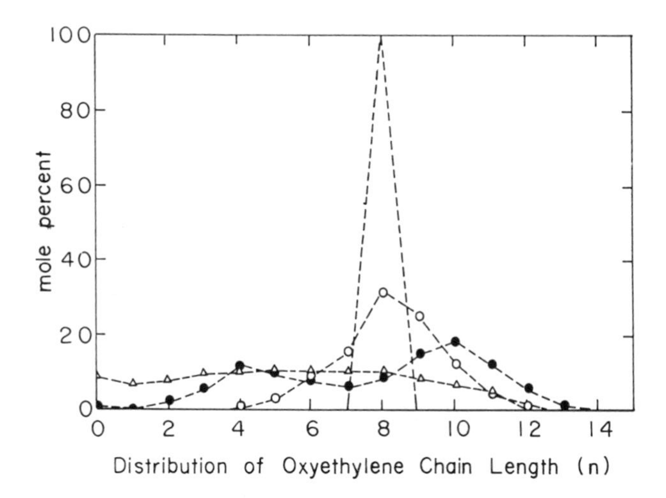

Figure 4.12 The distribution of the oxyethylene chain length of polyoxyethylene dodecylethers determined by programmed temperature gas chromatographic analysis. [---, homogeneous $R_{12}O(CH_2CH_2O)_8H$]; [-○-, molecular distilled $R_{12}O(CH_2CH_2O)_{8.2}H$]; [-●-, mixture of molecular distilled materials $R_{12}O(CH_2CH_2O)_{8.0}H$]; [-Δ-, commercial $R_{12}O(CH_2CH_2O)_{5.3}H$]. [Reproduced by permission of Academic, New York, *J. Colloid Interface Sci.*, **35**, 624 (1971).]

of the polyoxyethylene chain in surfactants of the alkylarylether type is represented by Poisson's distribution (19). This is due to the fact that the alkylphenol hydrogen is more reactive than the resulting alkylphenoxyethanol hydrogen. The distribution of the polyoxyethylene chain was calculated assuming Poisson's distribution and plotted in Fig. 4.13 (14).

Sample 1, purified by solvent (1 : 2 mixture of c-C_6H_{12} and n-C_6H_{14}) extraction, showed a slight decrease in short and long oxyethylene chain compounds by gas chromatographic analysis, but the shape of the distribution scarcely differs from that of sample 1. It is evident from Table 4.5 and Fig. 4.13 that the stability of an emulsion increases when the hy-

TABLE 4.5 The Effect of the Distribution of the Hydrophilic Chain Length of Polyoxyethylene Nonylphenylether of the Same PIT on the Stability of O/W-Type Emulsions[a]

Emulsifier $R_9C_6H_4O(CH_2CH_2O)_{\bar{n}}H$	PIT (°C)	Mean Droplet Diameter (μ)		Volume Percent after 60 days		Relative Stability for Coalescence
		0 day	60 days	Drained Phase	Coalesced Phase	
1. $\bar{n}$ = 8.6 [purified from the commercial sample (2)]	60.3	1.4	4.0	40	3	0.7
2. $\bar{n}$ = 8.6 (commercial)	60.5	1.6	3.5	40	3	1 (standard)
3. $\bar{n}$ = 8.2 (mixture) $\bar{n}$ = 9.7 38 wt% $\bar{n}$ = 7.4 62 wt%	60.5	1.4	3.3	40	3	1
4. $\bar{n}$ = 7.7 (mixture) $\bar{n}$ = 11.0 40 wt% $\bar{n}$ = 6.2 60 wt%	61.0	1.4	2.7	40	2.5	2
5. $\bar{n}$ = 7.5 (mixture) $\bar{n}$ = 14.0 25 wt% $\bar{n}$ = 6.2 75 wt%	60.0	1.6	2.6	40	2.5	2

[a]The emulsions are composed of 48.5 wt% of water, 48.5 wt% of cyclohexane, and 3 wt% of nonionic emulsifiers. The systems were shaken at 59°C, cooled, and stored at 25°C over 60 days.

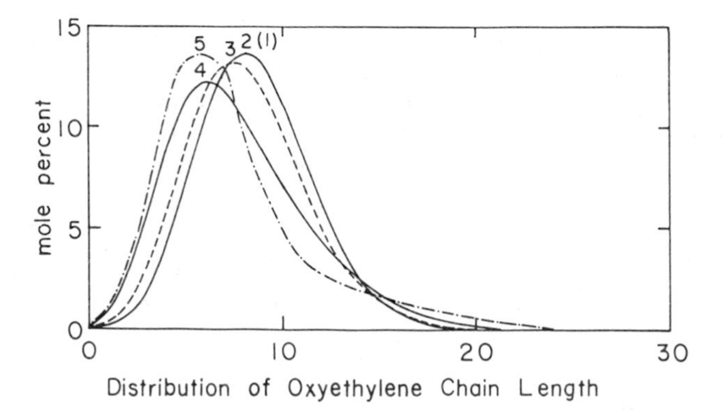

Figure 4.13 The distribution of the hydrophilic chain length of polyoxyethylene nonylphenylethers of the same PIT (60°C), calculated based on a Poisson distribution. For the key to the numbers on the figure, see Table 4.5. [Reproduced by permission of Academic, New York, *J. Colloid Interface Sci.*, **35**, 624 (1971).]

drophilic groups are fairly broadly distributed. However, the optimum broadness was not observed within this experiment.

The PIT becomes ambiguous and changes sensitively with concentration when an emulsifier is composed of a mixture of surfactants of widely different HLB numbers. Thus, various mixtures of polyoxyethylene nonylphenylethers, the average oxyethylene chain lengths of which are 8.5 were prepared, and the stability of emulsions was observed for over 2 years. The results are summarized in Table 4.6 and the distributions of the polyoxyethylene chains are plotted in Fig. 4.14.

It can be concluded from Tables 4.5 and 4.6 that the mixing of long and short oxyethylene chain homologs is effective in increasing the stability of emulsions in the case of polyoxyethylene alkylarylethers. In the case of polyoxyethylene alkylether synthesized with alkali catalyzer, the distribution of

TABLE 4.6 The Effect of the Distribution of the Hydrophilic Chain Length of Polyoxyethylene Nonylphenylether of the Same HLB number on the Stability of O/W-Type Emulsions[a]

Emulsifier H $R_9C_6H_4O(CH_2CH_2O)_{8.6}$	Mean Droplet Diameter (μ) 20 months	Volume Percent after 20 months: Drained Phase	Volume Percent after 20 months: Coalesced Phase
1. $\bar{n} = 8.6$ (commercial)	20	40	4
2. Mixture	15–20	40	2.5
$\bar{n} = 9.7$ (0.5 mole)			
$\bar{n} = 7.4$ (0.5 mole)			
3. Mixture	15–20	40	2.5
$\bar{n} = 11$ (0.5 mole)			
$\bar{n} = 6.2$ (0.5 mole)			
4. Mixture	15–20	40	2.5
$\bar{n} = 14$ (0.3 mole)			
$\bar{n} = 8.6$ (0.4 mole)			
$\bar{n} = 3$ (0.3 mole)			
5. Mixture	20	40	3
$\bar{n} = 14$ (0.4 mole)			
$\bar{n} = 8.6$ (0.2 mole)			
$\bar{n} = 3$ (0.4 mole)			
6. Mixture	40	40	4
$\bar{n} = 17.7$ (0.37 mole)			
$\bar{n} = 3$ (0.63 mole)			

[a]The emulsions are composed of 48.5 wt% of water, 48.5 wt% of cyclohexane, and 3 wt% of nonionic surfactant. The systems were shaken and stored at room temperature (10–35°C) over 20 months.

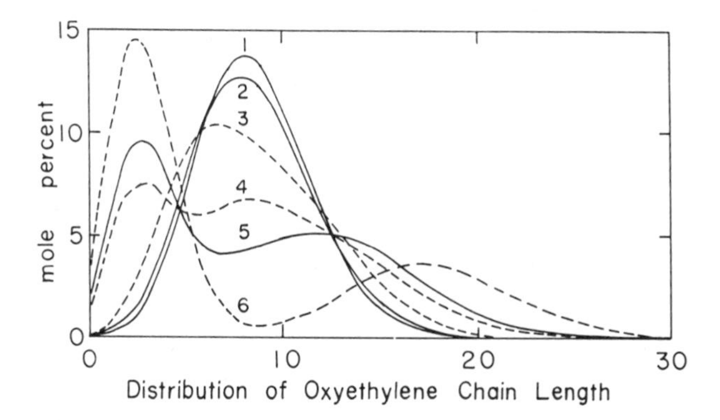

Figure 4.14 The calculated distribution of polyoxyethylene chain lengths of $R_9C_6H_4O(CH_2CH_2O)_{8.6}H$, based on a Poisson distribution. For the key to the numbers on the figure see Table 4.6. [Reproduced by permission of Academic, New York, *J. Colloid Interface Sci.*, 35, 624 (1971).]

the oxyethylene chain is fairly wide so that mixing may not be effective.

4.3.3 The Effect of the Substitution of $C_9H_{19}C_6H_4O(CH_2CH_2O)_{8.6}H$ with $C_{12}H_{25} \cdot C_6H_4SO_3Ca_{1/2}$

This type of mixture is widely used for the formulation of emulsifiers. Calcium salts are not hydrophilic and show a low concentration for the primary association in nonaqueous media. The emulsion stability of the mixtures is shown in Table 4.7.

The size of emulsion droplets becomes smaller and the stability for coalescence and drainage increases with the amount of the minor component of the mixture. The rough ratio of stability regarding drainage and coalescence are also

TABLE 4.7 The Effect of the Substitution of $R_9C_6H_4O(CH_2CH_2O)_{8.6}H$ with $R_{12}C_6H_4SO_3Ca_{1/2}$ on the Stability of O/W-Type Emulsions[a]

Mixed Emulsifiers	(wt %)	PIT (°C)	Emulsification Temperature (°C)	Mean Droplet Diameter (μ)		Volume percent after 20 days		Relative Stability for Coalescence
				0 day	20 days	Drained Phase	Coalesced Phase	
$R_9C_6H_4O(CH_2CH_2O)_{8.6}H$	100 %	60.5	59	1.6	3.0	40	1	1 (standard)
$R_9C_6H_4O(CH_2CH_2O)_{8.6}H$	95 %	>100	80[b]	1.6	2.2	32	1	8
$R_{12}C_6H_4SO_3Ca_{1/2}$	5 %							
$R_9C_6H_4O(CH_2CH_2O)_{8.6}H$	90 %	100	76[b]	1.5	2.2	29	1	12
$R_{12}C_6H_4SO_3Ca_{1/2}$	10 %							
$R_9C_6H_4O(CH_2CH_2O)_{8.6}H$	80 %	100	66[b]	1.4	1.5	18	1	50
$R_{12}C_6H_4SO_3Ca_{1/2}$	20 %							

[a]The emulsions are composed of 48.5 wt % of water, 48.5 wt % of cyclohexane, and 3 wt % of emulsifiers. The systems were shaken at the respective emulsification temperature indicated, cooled, and stored at 25°C over 20 days.

[b]The optimum temperature for emulsification.

shown in Table 4.7. These ratios correspond to the time required to pass the same stage for each system compared with the original system.

REFERENCES

1. K. Shinoda and H. Saito, *J. Colloid Interface Sci.*, **26**, 70 (1968).
2. K. Shinoda and H. Saito, *J. Colloid Interface Sci.*, **30**, 258 (1969).
3. H. Saito and K. Shinoda, *J. Colloid Interface Sci.*, **32**, 647 (1970).
4. K. Mandani and S. Friberg, *Progr. Colloid Polym. Sci.*, **65**, 164 (1978).
5. H. Kunieda and K. Shinoda, *Bull. Chem. Soc. Jpn.*, **55**, 1777 (1982).
6. S. Friberg and C. Solans, *J. Colloid Interface Sci.*, **66**, 367 (1978).
7. K. Shinoda and H. Sagitani, *J. Colloid Interface Sci.*, **64**, 68 (1978).
8. K. Shinoda and H. Kunieda, *J. Colloid Interface Sci.*, **42**, 381 (1973).
9. W. C. Griffin, *J. Soc. Cosmet. Chem.*, **1**, 311 (1949).
10. W. C. Griffin, *J. Soc. Cosmet. Chem.*, **5**, 249 (1954).
11. Atlas Chemical Industries, Inc., *The Atlas HLB System*, Wilmington, Delaware (1963).
12. P. Becher, *Emulsions*, 2nd ed., Reinhold, New York (1966).
13. Figure 6b in Ref. 2.
14. K. Shinoda, H. Saito, and H. Arai, *J. Colloid Interface Sci.*, **35**, 624 (1971).

15. R. L. Mayhew and R. C. Hyatt, *J. Am. Oil Chem. Soc.*, **29**, 357 (1952).
16. E. H. Crook and D. B. Fordyce, *J. Am. Oil Chem. Soc.*, **41**, 231 (1964).
17. K. Shinoda and H. Arai, *J. Phys. Chem.*, **68**, 3485 (1964).
18. K. Shinoda, *J. Colloid Interface Sci.*, **25**, 429 (1967).
19. K. Konishi and S. Yamaguchi, *Anal. Chem.*, **38**, 1755 (1966).
20. K. Shinoda (Ed.), in *Solvent Properties of Surfactant Solutions*, Chapter 2, Marcel Dekker, New York (1967).
21. J. H. Schulman and E. G. Cockbain, *Trans. Faraday Soc.*, **36**, 651 (1940).
22. E. H. Crook, D. B. Fordyce, and G. F. Trebbi, *J. Phys. Chem.*, **67**, 1987 (1963).

CHAPTER 5

Liquid Crystals and Emulsions

The discussion about the surfactant phase (1–5) shows this irregular, optically isotropic liquid structure to be obtained when the solubility of the water and the hydrocarbon in the surfactant is balanced. This means that the radius of curvature for the structure is zero and irregular structures of the kind suggested by Larson and Scriven (6) are found. With higher surfactant concentrations the layers become ordered and a liquid crystalline phase is obtained. Figure 5.1 illustrates the concentration dependence of this structure change. This phase transition with increased concentration is similar to the one found in monomolecular layers (7–9) for which

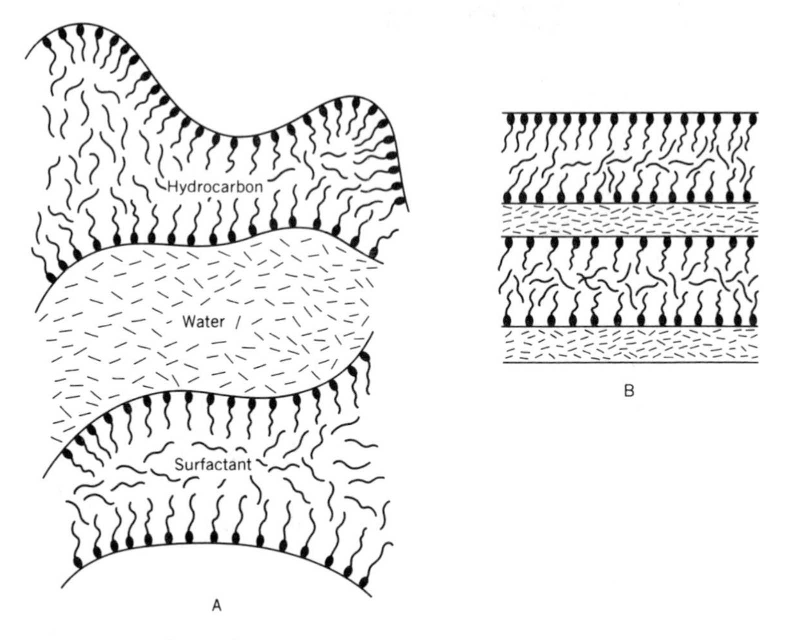

Figure 5.1 The surfactant phase contains aggregates with an averaged radius of curvature equal to zero in a liquid isotropic form, A. Higher concentrations of the surfactant gives rise to a lamellar close-packed structure in which both the principal curvatures are zero for all sites, except where dislocations occur, B.

the transition to a more close-packed phase is encountered with reduced area per molecule.

This similarity may be brought even further. In the monomolecular layer a long chain carboxylic acid such as stearic acid does not form a liquid expended structure. Instead, a more close-packed structure is directly formed from the gaseous state, when the area per molecule is reduced.

Analogous behavior is found in systems with nonionic surfactants if the polar chain is of sufficient length. No isotropic liquid surfactant phase is found; instead, a lamellar liquid crystal is separated at low concentrations of surfactant.

An illustrative example of such behavior is found in the system water, *p*-xylene, and an industrial octaethyleneglycol nonylphenolether, Fig. 5.2 (10). In this case, a lamellar liquid crystalline phase is separated at surfactant concentrations as low as 3% by weight of the total, which may be related to the fact that the solubility of aromatic hydrocarbon and water in the nonionic surfactant is large. The presence of this phase will give an enhanced stabilization to the emulsion (11–14) and also change its rheological properties (15–17).

Although it is illustrative to separate the liquid crystalline phase from the emulsion, the presence of it in the emulsion may be determined directly in the microscope under polarized light. The lamellar or hexagonal liquid crystals are optically anisotropic and its particles or layers around the droplets are conspicuous features in the microscope. Figure 5.3 shows an emulsion that contains droplets with liquid crystals, as well as droplets without them. The droplets with liquid crystals are a conspicuous feature; they show a strong halo in polarized light.

A separation of the phases in the emulsion makes it possible to discover the presence of a liquid crystalline phase also without the use of a microscope. The liquid crystalline layer

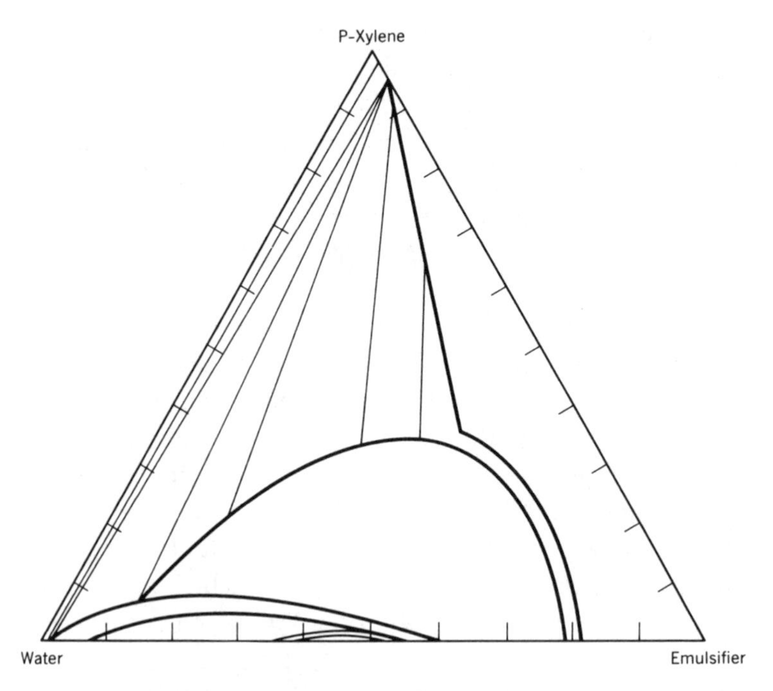

Figure 5.2 In the system water, *p*-xylene, and an industrial emulsifier (approximately octaethyleneglycol nonylphenolether) no surfactant phase is found at room temperature. Instead a regular lamellar liquid crystal is separated at small concentrations of the surfactant.

will appear radiant, when viewed against a light source through crossed polarizers, Fig. 5.4. This separation is usually achieved by centrifugation and a note of caution is justified in this context. Large colloidal particles such as those found in the surfactant phase, Fig. 5.1, move in the strong field of an ultracentrifuge. Ultracentrifuging the surfactant phase results in three phases of which one is a liquid crystalline phase. This is not the stable state however, if left with only the gravity field acting on the three phases they will spontaneously revert to the surfactant phase; the stable one.

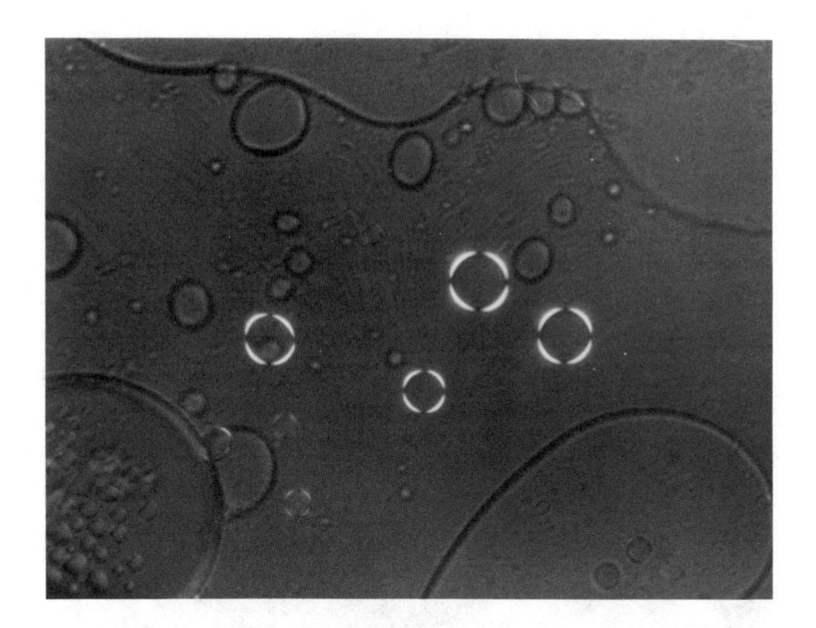

Figure 5.3 The droplets covered with a liquid crystal with their halos and Malthesian crosses are a conspicuous feature in an emulsion.

Figure 5.4 When the three phases in an emulsion containing liquid crystals have been separated the liquid crystalline layer is identified because it is radiant, when viewed between crossed polarizers against a light source. The two liquid phases are optically isotropic and appear dark.

When a separation has been achieved, the structure of the separated liquid crystalline phase can be determined by the use of the optical microscope. The lamellar structure will show the "neat" pattern, and/or the spherical patterns with the Malthesian crosses, Fig. 5.5. The liquid crystalline structure of hexagonally packed cylinders will give the optical pattern in Fig. 5.6 (18).

These structures may be confirmed by using x-ray diffractometry. Figure 5.7 shows a comparison between the crystalline and the liquid crystalline lamellar pattern. The crystalline pattern is characterized by several sharp lines in the 3.5–4.6-Å range while the liquid crystalline pattern displays one diffuse reflection at 4.5 Å. The specific structure of the liquid

Figure 5.5 The microphoto pattern of a lamellar liquid crystal in polarized light shows both radiant bonds "oily streaks" and Malthesian crosses.

Figure 5.6 The microphoto pattern of an hexagonal liquid crystal in polarized light shows a pattern characterized by fanlike textures.

crystal is determined from the pattern in the low angle region (19). The lamellar structure shows a series of reflections with spacing ratios $1:2:3:4\cdots$ while the hexagonal structure displays the corresponding ratios $1:\sqrt{4}:\sqrt{7}$. An isotropic liquid crystalline structure shows no optical pattern in polarized light and hence can only be identified from its x-ray patterns.

The liquid crystals enhance the stability of the emulsion due to the fact that they form a covering skin around the droplets and also form a rigid three-dimensional network through the continuous phase. The covering skin prevents coalescence of droplets because of its high viscosity and the fact that a layered structure changes the distance depen-

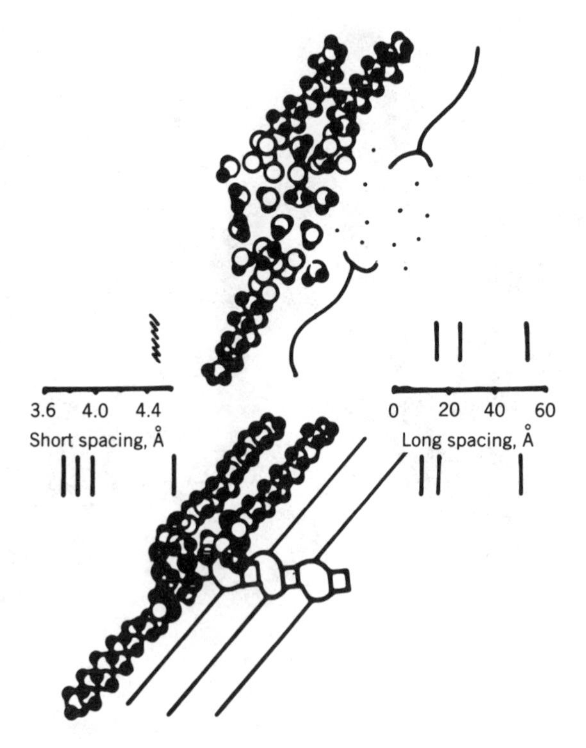

Figure 5.7 The crystalline state of a monoglyceride (lower part) is characterized by a short-range order both in the hydrocarbon chains and the polar parts (sharp reflections in the short spacing region, left). Addition of water (upper part) destroys the short-range order (diffuse reflection in the short spacing region, left); the long-range order prevails (long spacing, right) although the distance is changed.

dence of the van der Waals potential (20). In addition, the presence of charged molecules in the liquid crystalline phase will improve the stability by the influence of the electric double layer (21).

In order to understand the proper mixture of surfactants to form liquid crystals, the approach by Israelachvili et al.

and Mitchell and Ninham (22,23) is useful. The lamellar liquid crystals form when the ratio

$$R = \frac{v_H}{l_c a_0}$$

is in the range 0.5–1.0.

v_H = volume of hydrocarbon chain
l_c = length of hydrocarbon chain
a_0 = cross-section area of polar group

A typical representative of surfactants to form such a structure is lecithin, other possibilities include monoglycerides (24), alkylammonium alkylcarboxylates (12), and the combination long chain alcohol/long chain soap (18), long chain acid/long chain soap or long chain soap/polyethyleneglycol dodecylether with short polar chain (25).

Emulsions with liquid crystals show excellent stability and high viscosity and exist, *nolens volens*, in many cosmetic formulations. They are, of course, no panacea, in order to prepare stable emulsions; emulsions with excellent stability may be easily prepared in a two-phase system.

However, the liquid crystalline phase has one characteristic that may turn out to give these emulsions a specific advantage in formulation efforts. The liquid crystalline structure may be used to dissolve substances that otherwise show only limited solubility. A good example is hydrocortisone, the solubility of which in isotropic solvents is small; approximately 1.5% in ethylene glycol. A lamellar liquid crystalline phase of lecithin and water dissolved more than 4% of the substance (26). It appears obvious that emulsions with liquid crystalline phases may find use for specific formulations in which the active substance is soluble only in the liquid crystalline phase.

Another property of importance is the low diffusion coefficient of solubilized substances (16). The emulsion with liquid crystalline phases has shown interesting properties as vesicles for delayed delivery systems for drugs. Sylvan Frank (27,28) is the leading researcher in this area and has made extensive investigations.

REFERENCES

1. K. Shinoda and H. Saito, *J. Colloid Interface Sci.*, **26**, 70 (1968).
2. K. Shinoda and H. Kunieda, *J. Colloid Interface Sci.*, **42**, 381 (1973).
3. K. Shinoda and S. Friberg, *Adv. Colloid Interface Sci.*, **4**, 281 (1975).
4. K. Shinoda, H. Kunieda, N. Obi, and S. E. Friberg, *J. Colloid Interface Sci.*, **80**, 304 (1981).
5. H. Kunieda and S. E. Friberg, *Bull. Chem. Soc. Jpn.*, **54**, 1010 (1981).
6. R. G. Larson, L. E. Scriven, and H. T. Davis, *Nature*, **268**, 409 (1977).
7. H. C. Parreira, *J. Colloid Interface Sci.*, **20**, 742 (1965).
8. N. L. Gershfeld and Y. G. Pak, *J. Colloid Interface Sci.*, **23**, 215 (1967).
9. H. H. Bruun, *Acta Chem. Scand.*, **9**, 1721 (1955).
10. S. Friberg, L. Mandell, and M. Larsson, *J. Colloid Interface Sci.*, **29**, 155 (1969).
11. S. Friberg and L. Mandell, *J. Am. Oil Chem. Soc.*, **47**, 149 (1970).
12. S. Friberg, *Kolloid Z.Z. Polym.*, **244**, 333 (1971).

13. S. Friberg, P. O. Jansson, and E. Cederberg, *J. Colloid Interface Sci.*, **55**, 614 (1976).
14. S. Friberg and K. Larsson, *Advances in Liquid Crystals*, Vol. 2, Academic, New York (1976), p. 173.
15. S. Friberg and P. Solyom, *Kolloid Z.Z. Polym.*, **236**, 173 (1970).
16. B. W. Barry and G. M. Eccleston, *J. Texture Stud.*, **4**, 53 (1973).
17. B. W. Barry, *Adv. Colloid Interface Sci.*, **5**, 37 (1975).
18. P. Ekwall, in *Advances in Liquid Crystals* (G. Brown, Ed.), Vol. 1, Academic, New York (1975), p. 1.
19. K. Fontell, *Mol. Cryst. Liq. Cryst.*, **63**, 59 (1981).
20. P. Jansson and S. Friberg, *Mol. Cryst. Liq. Cryst.*, **34**, 75 (1976).
21. L. Rydhag and I. Wilton, *J. Am. Oil Chem. Soc.*, **58**, 830 (1981).
22. J. Israelachvili, D. J. Mitchell, and B. W. Ninham, *J. Chem. Soc. Faraday Trans. II*, **72**, 1525 (1976).
23. D. J. Mitchell and B. W. Ninham, *J. Chem. Soc. Faraday Trans. II*, **77**, 601 (1981).
24. K. Larsson, in *Food Emulsions* (S. E. Friberg, Ed.) Marcel Dekker, New York (1976), pp. 39–66.
25. H. Sagitani and S. Friberg, *Bull. Chem. Soc. Jpn.*, **56**, 31 (1983).
26. S. Wahlgren, A. Lindstrom, and S. Friberg, *J. Pharm. Sci.* (in press 1985).
27. S. M. Ng and S. G. Frank, *J. Dispersion Sci. Technol.*, **3(3)**, 271 (1982).
28. D. R. Kavaliunas and S. G. Frank, *J. Colloid Interface Sci.*, **66**, 586 (1978).

Index

Alkane, effect of chain length on PIT, 108

Bending surface energy, 13, 33, 160
Benzene, 22, 64, 65, 86, 97, 102, 107, 109
Bromobenzene, 106–107

Carbon tetrachloride, 65, 86, 98
Cloud point:
- of aqueous surfactant solution, 60
- of surfactant solution:
  - with solubilized oil, 19–40, 60, 105, 118
  - with solubilized water, 23, 28–40, 105, 118

Coalescence, 126–135, 142
- increase of stability for, 155–156
- of O/W type emulsion, 146
- of W/O type emulsion, 147

Competitive dissolution, of water and oil in surfactant phase, 32–34, 42
Concave or convex curvature, of adsorbed monolayer, 13, 14, 21–22, 24, 33, 62
Conceptual diagram, of dialkyl type ionic surfactant, 90
Contour, stability of emulsions, 137, 139–141
Critical solution, of surfactant-oil or water, 41–42, 49
Crossed polarizer, 162
Cyclohexane, 22–23, 26, 30, 35–40, 59, 63–65, 88–90, 97, 99, 102–106, 109–111, 118, 121, 127–135, 137–141, 146–147, 150–156

Decane, 22, 48
Definition:
- of emulsion, 3
- of HLB, HLB number, HLB temperature, 6
- of microemulsion, 5
- of solubilization, 4

Detergency, optimum temperature, 34
Dialkyl type surfactant, 41, 89–90
Dispersion, of spilled oil, 121
Dispersion type, 37
- hydrophilic chain length of nonionics, effect of, 34–40
- temperature, effect of, 36–40

Dodecene (propylene tetramer), 65, 86, 98, 107
Drug, delayed delivery system, 168
Dry cleaning, 29

Electron micrograph, 5
Emulsification, by PIT method, 129–131, 135, 145
Emulsifier:
  of same average hydrophilic chain length and different distribution, 144, 148–155
  of same HLB number but different ethyleneoxide distribution, 144, 148–155
  of same PIT(HLB) and different size, 144–148
Emulsion, definition of, 3
Emulsion inversion point(EIP), 56
Emulsion type:
  solution behavior of surfactant, change with, 12–50
  temperature, change with, 24–29, 36–40, 96–106
Ethylbenzene, 22, 98
Ethyl oleate, 99

Glycerol monoether, as cosurfactant, 42–49

Haze point, 23, 118
Heptane, 20, 22, 64, 97, 99, 102, 108, 110–113, 120
Hexadecane, 22, 64, 97, 143
HLB (hydrophile-lipophile balance, hydrophilic-lipophilic balance):
  composition:
    change with, 36, 41–49, 56
    at which hydrophile-lipophile property balances, 42–47, 56, 89
  concept of, 56–58
  definition of, 6
  of ionic surfactant(control), 40–49, 88–90
  temperature, change with, 19–40, 63–64, 121, 127–142
HLB group number, 83–84
HLB number:
  definition of, 6
  HLB temperature (PIT), comparison with, 58–68, 102, 121
  PIT, determination from, 63–65, 102, 113, 121
  required:
    of oil, 60, 63–65, 140–143
    of surfactant, 58–59
HLB system, 58
HLB temperature:
  concept of, 58–67
  definition of, 6
  dissolution, characteristic temperature for, 32–34
  emulsion, characteristic temperature for, 57, 60
  HLB number, comparison with, 58–67, 121
  see also PIT
H/L number, 56, 71–73
Hydrophile-lipophile balance(HLB), *see* HLB (hydrophile-lipophile balance, hydrophilic-lipophilic balance)

Illustration:
  close-packed emulsion, 27
  surfactant, solution behavior of, 16–17, 33, 138
  swollen surfactant phase, structure of, 21, 160
Interfacial tension, minimum at HLB temperature, 12, 33, 127–128, 130
Ionic surfactant:
  calcium salt soluble in hard water, 44–47
  control of HLB, 40–47, 88–90
  HLB of, 88–91
  solution behavior of, 40–50

Krafft point, depression by addition of cosurfactant, 46–47

Light scattering, droplet size determination, 5, 20

Liquid crystal, 17, 25
relation with emulsion, 160–168
Liquid paraffin:
cloud point in presence of, 22
PIT, 64–65, 98, 100, 102, 109, 114–115, 122
solubilization of, 46, 49
stability of emulsion of, 140

Malthesian cross, 163
Methyl naphthalene, 98
Microemulsion:
definition of, 5
diameter of, 5
solubilized solution, 35–36, 43–49
solution behavior of surfactant, correlation with, 12–50
see also Solubilization
Microscopic photograph:
of emulsion, 135
of liquid crystal in emulsion, 163–165
Monoglyceride, crystalline state of, 166

Perchloroethylene, 22, 99, 110
Phase equilibria:
added salt, effect of, 118
hydrophilic chain length, influence of, 34–36
ionic surfactant system, 40–50
temperature, effect of, 28–34, 36–40
Phase inversion temperature in emulsion, *see* HLB temperature; PIT
PIT:
added salts, acid, alkali, effect of, 117–120
emulsification by method, 129–135
emulsifier:
concentration, effect of, 97–101, 105
hydrophilic chain length of, effect of, 99–102
mixed, effect of, 110–117
molecular formula of, 66–68, 100–103, 113–115, 121, 140–141
oxyethylene chain length of, 99–102
emulsion:
characteristic property of, 57, 60
stability, 59, 126–129, 133–135, 137–141
factors affecting, 96–122
HLB number, correlation with, 61–67, 121, 140–143
oil:
additives in, effect of, 122
hydrocarbon chain length of, effect of, 108
mixed, effect of, 108–110
types of, effect of, 96–100, 102, 107
phase volume, effect of, 26–28, 98–100, 103–107
see also HLB temperature
Polymethylphenylsiloxane, 100

Sedimentation velocity, 5
Sodium pyroglutamate, 43
Solubility curves:
among water, ionics, cosurfactant, oil, 43–46, 48–49
among water, nonionics, hydrocarbon, 20, 23, 30, 35–37, 39, 118
Solubilization:
definition of, 4
of oil, 20, 30, 35–37, 39, 118
of water, 23, 30, 35–37, 39, 118
see also Microemulsion
Span, 74–78, 85, 112–117
Squalane, 100, 102
Stability of emulsion:
emulsification temperature, effect of, 126–127, 136–141
emulsifier size, effect of, 144–148
HLB number, insensitive to, 128–134
hydrophilic chain length, effect of, 128–129, 138–143, 150–155
liquid crystal, stabilization by, 162, 167

Stability of emulsion (*Continued*)
oxyethylene chain, distribution of, effect of, 144, 148–155
PIT:
as function of, 128–129, 132–135
sensitive to, 128–134
raising temperature as invalid test of, 146–149
after twenty months, 154
W/O type, 127, 136–144, 147
Swollen surfactant phase:
lamellar liquid crystal, similarity with, 160
separation of, 15, 19–20
structure of, 21, 33, 160
three phase region, 35–40, 43–46, 138

Three phase region:
between two critical points, 49
of water, surfactant, oil phases, 33, 35–40, 42, 118, 138
Toluene, 65, 86, 97
Trichloro-trifluoroethane, 65, 99
Tween, 78–81, 85, 112–117

X-ray scattering, 164–165
small angle, 5
Xylene, 64–65, 86, 97, 102, 107, 109, 162